The Limits of Science

Poznań Studies in the Philosophy of the Sciences and the Humanities

VOLUME 109

The titles published in this series are listed at *brill.com/ps*

The Limits of Science

An Analysis from "Barriers" to "Confines"

Edited by

Wenceslao J. Gonzalez

BRILL
RODOPI

LEIDEN | BOSTON

Poznan Studies is sponsored by the University of Warsaw.

Cover illustration: © Jessica Rey.

Library of Congress Cataloging-in-Publication Data

Names: Gonzalez, Wenceslao J., editor.
Title: The limits of science : an analysis from "barriers" to "confines" / edited by Wenceslao J. Gonzalez.
Description: Leiden ; Boston : Brill-Rodopi, 2016. | Series: Poznań studies in the philosophy of the scienc and the humanities, ISSN 0303-8157 ; VOLUME 109 | Includes bibliographical references and index.
Identifiers: LCCN 2016028031 (print) | LCCN 2016028909 (ebook) | ISBN 9789004325395 (hardback : alk. paper) | ISBN 9789004325401 (E-book)
Subjects: LCSH: Science--Philosophy.
Classification: LCC Q175 .L543 2016 (print) | LCC Q175 (ebook) | DDC 501--dc23
LC record available at https://lccn.loc.gov/2016028031

Want or need Open Access? Brill Open offers you the choice to make your research freely accessible onli in exchange for a publication charge. Review your various options on brill.com/brillopen.

Typeface for the Latin, Greek, and Cyrillic scripts: "Brill". See and download: brill.com/brill-typeface.

ISSN 0303-8157
ISBN 978-90-04-32539-5 (hardback)
ISBN 978-90-04-32540-1 (e-book)

This book is printed on acid-free paper and produced in a sustainable manner.

In Memory of Antonio Bereijo

Contents

Preface: The Problem of the Limits of Science in the Present Context

Wenceslao J. Gonzalez

Initially, the problem of limits of science is twofold. On the one hand, it deals with the boundaries between what is science and what is not science, either because it has not enough theoretical and empirical support to be considered "science" now, or because it is an undertaking outside the "barriers" or frontiers of the territory of what is accepted as scientific endeavor. On the other hand, the problem of limits of science also discusses the ceiling of scientific activity, which leads to the characterization of the "confines" of this human enterprise. This reflection includes considering if there is a point or a realm that is beyond science, either because science is unable to reach it now, or because such point or realm cannot be got at any time in the foreseeable future.

These two faces of the problem of the limits – the "barriers" or frontiers and the "confines" or ceiling of science – require a new analysis, which is the task of this book. Thus, it takes into account the Kantian roots, the distinction between both kinds of limits: *Schranken* (barriers) and *Grenzen* (confines). In this regard, it considers this twofold approach to limits of science according to the present stage of the philosophico-methodological analysis of science. This vision that follows the present context is well aware of the historicity of science. Subsequently, it looks to supersede the Kantian mold of science as a content in order to reach the richer conception of science as a human activity developed in a social milieu.

1 Aspects to be Considered in the Problem of Limits of Science

Following the vision of science of the present context, the problem of limits of science includes a set of aspects to be considered. First, the analysis should be made according to what we nowadays think of what science *is* and *ought to be*. This approach involves the possibility of a study of the limits concerning each component of a science. Thus, the focus of the analysis can be on the language, structure, knowledge, methods, activity, ends, and values of science. Consequently, the problem of the limits of science can be analyzed philosophically from these constitutive elements of a science. This leads to a variety of possible analyses of the limits of science: semantic, logical, epistemological, methodological, ontological, axiological, and ethical.

Second, there are several possible *levels of analysis* of the limits according to the degree of generality of the philosophico-methodological scope. The main levels are three: (i) science in general, when it is focused on the features valid, in principle, for any science (or, at least, for any empirical science); (ii) a group of sciences, when the analysis looks for characteristics present in the natural sciences, the social sciences, or the sciences of the artificial; and (iii) specific sciences, when the philosophico-methodological consideration deals with traits that belong to specific disciplines (physics, biology, economics, sociology, information science, computer sciences, etc.).

Third, the problem of the limits of science can also have some features on account of the *kind of research*. In this regard, there are again three main options: basic science, applied science, and the application of science. When the philosophico-methodological analysis deals with basic science, which is related to the central aim of the rigorous advancement of knowledge, the limits of science have to do with explanation and prediction. When, however, the philosophical consideration discusses applied science, which is focused on the solution of concrete problems within a domain, the limits of science pay attention to the issues related to prediction and prescription. Thereafter, the philosophical study of the application of science, which is the use of scientific knowledge by the agents within the practical contexts of the professions, requires taking care of the limits of science according to the mediation of the changeable contexts (social, cultural, economic, etc.). All of these include the historicity of the scientific activity.

Fourth, science is a human undertaking that involves *structural elements* and *dynamic components*, which are more noticeable when we are considering a complex system. The presence of complexity – in science, in general, in a group of sciences, or in a particular discipline – adds new facets to the problem of the limits. An analysis of these complex ingredients in terms of limits can be made in the realm of the structure or in the sphere of the dynamics. Thus, we can consider limits of science in four different ways at least: hological, etiological, teleological, and logical.

The holological analysis should offer us the limits of the decomposability (or near decomposability) of a complex system, which may be the limits of knowability of the system in question (natural, social, or artificial). The etiological analysis can present us the causes – if there are any – and the possible nexi between several causes, which can lead also to limits of knowability, especially in some complex systems (for example, when the research in economics is about the limits of predictability of some financial markets). The teleological analysis should exhibit the relations between ends and means, where also the case of limits of knowability may be. The logical analysis can make explicit what is conceptually impossible.

Fifth, science is a human activity with an *"internal" side* and an *"external" dimension*. On the one hand, science has content related to the set of its components (semantic, logical, epistemological, methodological, ontological, axiological, etc.). On the other, science has a social dimension connected to a changeable environment, because it is a historical undertaking in a diversity of contexts (cultural, economic, political, etc.). The Kantian tradition on the limits of science in terms of *Schranken* (barriers) and *Grenzen* (confines) is commonly focused on the "internal" side of science and thus usually pays little attention to the "external" dimension of science.

Consequently, the problem of demarcation – a key issue for thinkers under direct Kantian influence, such as Karl Popper – as well as the issue of the possibility of a "perfect science" – a central issue of interest in Nicholas Rescher – are considered from epistemological and methodological viewpoints. Meanwhile, the studies of science in its relation to technology and society are frequently oriented to the "external" dimension, and the problems of science are seen then in terms of limitations (social, cultural, political, ecological, etc.).

2 Presence Here of the Variety of Aspects of Limits of Science to be Considered

This book offers features related to these five lines of philosophico-methodological analysis of the limits of science. These varied of aspects of the limits of science are considered in some ways in this volume. In this regard, following the same order of the factors already mentioned, some considerations can be made along these five lines. This is seen in the four parts of the book, where – to some extent – the different kinds of analysis appear in one way or another, and there is sometimes an interviewing with other kinds of analysis.

(1) A frequent trait of the analysis made in this volume is scientific status. Thus, it includes reflections on what science *is* and *ought to be*. The study of the limits now and in the foreseeable future is made concerning each component of a science (language, structure, knowledge, methods, activity, ends, and values). Nevertheless, the reflection habitually puts a special emphasis on three of them: the epistemological, methodological, and ontological features.

(2) Certainly, the three possible levels of analysis of the limits of science according to the degree of generality (science, a group of sciences, and specific sciences) are considered in this book in a number of ways. The task includes

philosophical remarks made about the three cases of the degree of generality, including reflections on the natural sciences, the social sciences, and the sciences of the artificial.

(3) Undoubtedly, the three main kinds of research (basic science, applied science, and application of science) are also considered, with reflections on the problem of the limits of science according to the features of the three kinds of scientific research. In addition, there are explicit and implicit reflections on the relations between scientific limits and philosophical limits as well as between limits of scientific research and limits of technological research (e.g., when reflecting on computers).

(4) Commonly, the authors of the chapters assume that science is a human undertaking with structural elements and dynamic components. This recognition is more noticeable in some chapters (mainly, 1, 7, 8 and 10). These structural elements and dynamic components can be seen in complex systems, where the configuration and evolution of each system can take a large variety of forms. The book offers analyses of these aspects from the viewpoint of the limits of science.

(5) Even more explicit than that duality (structural and dynamic) in science is the acknowledgment in this volume that science is a human activity with an "internal" side and an "external" dimension. It seems clear that some papers emphasize that science has a content (semantic, logical, epistemological, methodological, ontological, axiological, etc.), whereas other papers are aware of the social dimension of science as a human activity (i.e., the relevance of the context of science as a historical undertaking in a diversity of settings).

De facto, the table of contents recognizes explicitly several of these philosophico-methodological aspects. This can be seen in the titles of the parts of the book as well as in the titles of several chapters and the sections therein So, in Part 1, entitled "Limits as Frontiers and as Confines," it seems clear tha the problem of the limits of science is considered from several angles.

Thus, in Chapter 1 – "Rethinking the Limits of Science: From the Difficulties Regarding the Frontiers to the Concern about the Confines" – Wenceslao J. Gonzalez (University of A Coruña) offers a new analysis of the main point at stake: (a) the twofold domain of frontiers and confines; (b) the discussion on the frontiers or "barriers" (*Schranken*) following the criteria according to components of a science; (c) the frequent interest in the confines or "ceiling (*Grenzen*), taking into account the limits of science and the unavailability of perfect science as well as the limits due to complexity; and (d) the relevance of novelty for this topic.

Meanwhile, in Chapter 2 – "The Uncertain Frontier between Scientific Theories and Metaphysical Research Programmes" – Juan Arana (University o

Seville) discusses the Popperian proposal regarding demarcation. This central point of Popper's philosophy of science is considered in terms of the frontiers or "barriers" between science and metaphysics. Thereafter, in Chapter 3 – "Cognitive Problems and Practical Limits: Computers and our Limitations," Nicholas Rescher (University of Pittsburgh) looks at the epistemological limits of science in terms of the confines or "ceiling" for our knowledge, according to the practical limits of the computation of the information available now and in the future.

Within Part 2, which is devoted to "Two Poles of Analysis: Language and Ethics," the attention goes to two elements of a science: its language (with a cognitive content) and its ethical values (endogenous and exogenous). In Chapter 4 – "Language and the Limits of Science" – Ladislav Kvasz (University of Prague) rethinks the Kantian contribution to this topic from the perspective of the *analytical and expressive boundaries of language* in science. Subsequently, in Chapter 5 – "Ethical Limits of Science, Especially Economics" – Gereon Wolters (University of Konstanz) considers the repercussion of ethical values in science (mainly exogenous values). He puts special interest on economics, which is a field that has received an enormous philosophico-methodological attention in recent years (above all, since the beginning of the ongoing economic crisis in 2007).

Part 3 is directly focused on scientific knowledge. Under the title of "Epistemological Limits to Science" there are three contributions. In Chapter 6, "Predicting and Knowability: The Problem of Future Knowledge," Nicholas Rescher (University of Pittsburgh) considers knowability of the future, which includes the issues of detail and generality, and examines its wider implications for the condition of human knowledge. Meanwhile, in Chapter 7, in her paper "The Limits of Future Knowledge: An Analysis of Nicholas Rescher's Epistemological Approach," Amanda Guillan (University of A Coruña) goes deeper into this vision of human knowledge and the diversity of aspects involved.

Antonio Bereijo (University of A Coruña) examines in Chapter 8, entitled "The Limits of Information Science," the epistemological problem in the case of a contemporary discipline: information science. He considers this discipline from the point of view of its boundaries as scientific endeavor (i.e., as a scientific rather than a non-scientific activity); from the perspective of its possible confines in the present context as well as in the future. He sees the limits of information science in connection with its own scientific status, which he characterizes as an applied science of design. In addition, he is aware of that, besides the *endogenous aspects* of this problem-solving task in an artificial domain (with aims, processes, and results), the limits of information science also depend on *exogenous features*, which are those elements (social, cultural, economic, political, ecological, etc.) that surround this scientific endeavor.

Finally, in Part 4 there is a clear interest in science as a human activity with an "internal" side and an "external" dimension. Thus, it analyzes "The Limits from Inside and the Limits from Outside". In Chapter 9, devoted to "Rescher and Gadamer: Two Complementary Views of the Limits of Sciences," Alfredo Marcos (University of Valladolid) makes a comparison between the vision of the pragmatic idealism of the former and the hermeneutical approach of the latter. The first approach is mainly from inside the science, while the second conception sees the limits of science from an exterior cultural point of view. In Chapter 10, entitled "The Obstacles to Scientific Prediction: An Analysis of the Limits of Predictability from the Ontology of Science," Amanda Guillan (University of A Coruña) pays attention to limits due to the phenomena themselves: anarchy, volatility, etc., which pose real limitations for the predictors.

3 Acknowledgements

Undoubtedly, the main recognition goes to Antonio Bereijo. This book is devoted to his memory. He passed away unexpectedly on February 26, 2014, when just 50 years old. He was Titular Professor of Information Science and a member of the research group of Philosophy and Methodology of the Sciences of the Artificial at the University of A Coruña. His main line of research was the philosophico-methodological approach to information science in order to give it a new epistemological status as applied science of design. With this new theoretical foundation of his discipline he made substantial contributions to this field.

Each year, Antonio Bereijo was a key supporter of the conferences on contemporary philosophy and methodology of science at the University of A Coruña, which have been held annually since 1996 at the Campus of Ferrol during March. His intense dedication to university activities was always recognized and has had a clear expression in the Master's degree that he organized on information science in the digital environment. Among his publications there is one that is complementary to the topics that he discusses in this volume: Bereijo, A., "The 'Category of Applied Science': An Analysis of its Justification from 'Information Science' as Design Science," in Gonzalez, W.J. (ed.), *Scientific Realism and Democratic Society: the Philosophy of Philip Kitcher*, Poznan Studies in the Philosophy of Science and the Humanities, Rodopi, Amsterdam, 2011, pp. 327–350.

Besides Antonio Bereijo, this book should recognize the contribution of Amanda Guillan to the editorial tasks. Her dedication to this project has been remarkable. In addition, the contribution to edition of this volume made by Jose F. Martinez-Solano and Jessica Rey should be explicitly acknowledged. Finally, my gratitude to the Centre for Philosophy of Natural and Social Science at the London School of Economics, where this book was completed.

London, 13. 7. 2015

Notes on Contributors

Juan Arana
is Professor of Philosophy at the University of Seville and a member of the Royal Academy of Moral and Political Sciences of Spain: jarana@us.es

Antonio Bereijo
was Titular Professor of Information Science at the University of A Coruña and a member the research group of "The Philosophy and Methodology of the Sciences of the Artificial" at the University of A Coruña: abereijo@udc.es

Wenceslao J. Gonzalez
is Professor of Logic and Philosophy of Science at the University of A Coruña, a member of the *International Academy for Philosophy of Sciences*, and a *Visiting Fellow* at the Center for Philosophy of Science at the University of Pittsburgh: wencglez@udc.es

Amanda Guillan
is a member of the research group of "The Philosophy and Methodology of the Sciences of the Artificial" at the University of A Coruña: a.guillan@udc.es

Ladislav Kvasz
is Professor at the Institute of Philosophy, Czech Academy of Sciences, Prague: ladislavkvasz@gmail.com

Alfredo Marcos
is Professor of Logic and Philosophy of Science at the University of Valladolid: amarcos@fyl.uva.es

Nicolas Rescher
is Distinguished Professor of Philosophy at the University of Pittsburgh and Co-chair of the Center for Philosophy of Science at the University of Pittsburgh: rescher@pitt.edu

Gereon Wolters
is Emeritus Professor of the Department of Philosophy at the University of Konstanz and Co-chair of the Steering Committee of the ESF Research Networking Programme "The Philosophy of Science in a European Perspective": gereon.wolters@uni-konstanz.de

PART 1

Limits as Frontiers and as Confines

Rethinking the Limits of Science: From the Difficulties Regarding the Frontiers to the Concern about the Confines[1]

Wenceslao J. Gonzalez

Abstract

New scientific developments, where scientific creativity frequently connects with technological innovation, and the philosophical contributions of recent decades are two reasons for rethinking the limits of science. Within these coordinates, this paper follows a path of several steps:

(1) The recognition of three main levels of analysis of science as well as the diversity of philosophico-methodological outlooks to study this topic of the limits of science. (2) Frontiers and confines as the twofold domain that is commonly involved in the discussion. This duality has Kantian roots and includes a central issue: the "positive" criteria (what science is or should be) and the "negative" criteria (what science is not now or cannot be in the future). (3) The discussion of the frontiers or "barriers", which requires to clarifying the need for criteria for the frontiers or "barriers" and, thereafter, proposes criteria according to the components of science. (4) The frequent interest in the confines or "ceiling", where the limits of science are related to the unavailability of a perfect science and to the limits due to complexity. (5) The relevance of novelty is taken into account as a key factor in rethinking the limits of science.

Keywords

Limits – science – frontiers – confines – scientific creativity – technological innovation – novelty

Nowadays, there is a need for rethinking the limits of science. This is because there are new scientific and philosophical factors at stake. On the one hand, there are relevant changes introduced by scientific progress in recent decades, both "internal" (language, structure, knowledge, method, activity, values, etc.) and "external" (social, cultural, political, economic, etc.). On the other, there have been novel developments in the philosophy of science, both in the

1 The final version of this paper was prepared at the London School of Economics (Centre for Philosophy of Natural and Social Sciences). I am grateful to John Worrall for the invitation.

 | DOI 10.1163/9789004325401_002

extension of the realms of discussion (ontological, axiological, ethical, etc.) and in the presence of the new topics for the philosophical discussion about scientific features.[2]

Up to now, the philosophical issues related with the limits of science, in general, and of empirical sciences, in particular, have been mainly in two directions. First, whether there is a *border* between making science and doing non-science. Thus, if there is such edge, where can the frontier or "barrier" between a scientific research and a non-scientific activity be located? Second, whether science is always open to the future or it can be closed (or be declared completely "mature," "accomplished," or "finished") in the future, because of "internal" or "external" reasons. This requires considering science either as an open enterprise regarding the future or accepting that it can have *boundaries* at a given moment.

Consequently, this deals with how the science of the future might be. This includes some issues regarding prediction: (a) on the kind of scientific discoveries possible, which involves the question of what kind of *aims* may be achieved; (b) on the characteristics of the possible *contents* included in these discoveries; and (c) on the *results* that can follow from these contents obtained through the research. Thus, there might be a prediction about the future science (characteristics and aims) as well as a prediction within science itself (contents and results).[3] In this regard, there is the question as to whether there might be some *insolubilia* (i.e., problems that human science will never be able to solve),[4] either because they are legitimately outside the scientific field or because the use of scientific means cannot solve them at any time in the future.

Thus, this philosophico-methodological setting on the limits of science has two main realms of discussion. On the one hand, there is the domain of the frontiers or "barriers" (*Schranken*) of science, which is relevant insofar as it can clarify the perimeter of scientific research, either basic or applied, and thereafter its scientific application. On the other, there is the sphere of the confines or "ceiling" (*Grenzen*) of scientific research, which is important for figuring out the human possibilities now and in the future. These aspects are of interest not only for philosophers of science and scientists,[5] but also for many other professionals as well as for the public in general.

2 Among the new topics to be discussed is, for example, Barwick (2013).

3 There are interesting reflections on these issues in Rescher (2012).

4 On this issue of insolubilia, see Rescher (1999b, Ch. 8, pp. 111–127). This book has three new chapters and other differences with regard to the original edition: Rescher (1984). For these reasons, they are listed in this chapter as different books.

5 The interest among scientists in this topic has a long tradition. See, for example, Cattell (1896); Planck (1949); Hammond (1975); Weisskopf (1977); and Metcalfe and Ramlogan (2005).

Within these coordinates, this paper follows a path of several steps. (1) The recognition of three main levels of analysis of science and the diversity of philosophico-methodological outlooks to study this topic of the limits of science. (2) A twofold domain is commonly involved in the discussion: frontiers and confines. This duality has Kantian roots (Kant 1787) and includes a central issue: the "positive" criteria (what science is or should be) and the "negative" criteria (what science is not now or cannot be in the future). (3) The discussion of the frontiers or "barriers," which requires clarifying the need for criteria for the frontiers or "barriers" and, thereafter, proposes criteria according to the components of science. (4) The frequent interest in the confines or "ceiling," where the limits of science are related to the unavailability of a perfect science and to the limits due to complexity. (5) The relevance of novelty is taken into account as a key factor in rethinking the limits of science.

1 Levels of Analysis of Science and Philosophico-Methodological Outlooks

If we look at the future, the need for rethinking the limits of science is clear for philosophers, insofar as the problem of the limits impinges on two important set of issues regarding the scientific undertaking. In this regard, both sides of the problem of the limits of science – frontiers or "barriers" and confines or "ceiling" – can be taken into account. But the novelty here is in thinking of the limits of science at three different *levels*: basic science, applied science, and application of science.[6] This triple level can be seen as an advancement, insofar as the focus of attention among philosophers of science has been mainly on the first level. Consequently, the interest has been in the limits for explanation and prediction in science.

In principle, it seems that basic science (explanation and prediction), applied science (prediction and prescription),[7] and application of science can be considered concerning the problem of limits. *Primá facie*, each level of analysis of science – basic, applied, and application – can involve two aspects: (a) the characterization of something as scientific instead of non-scientific, and (b) the possibility of reaching an upper limit for these human activities (in the three levels), a point beyond which is no longer possible to get anything scientifically done by human beings.

But the study cannot merely be focused on what is nowadays available in science (i.e., the present stage of affairs in scientific matters). The focus should

6 On this three philosophico-methodological levels, see Gonzalez (2013b, pp. 17–18).

7 This is particularly relevant in the social sciences, see Gonzalez (1998).

be twofold: the "structural" viewpoint and the "dynamic" perspective on sc ence. Both should be taken into account. This requisite comes from the exi tence of a scientific progress, which can be seen in a retrospective approac as well as in a prospective way.[8] The existence of *historicity* in science – i cluding conceptual revolutions (Gonzalez 2011b) – makes clear the need for new consideration of this problem of the limits of science. Moreover, scientif progress has an undeniable historical embedding, insofar as each historic moment can consider the stage of scientific progress already reached and th possibilities of scientific advancement open to the future.

According to the dynamic perspective, the philosopher of science shoul distinguish two main possibilities: (i) the *contextual limits* of science, whic are the limits due to a historical situation in the development of science, an (ii) the *intrinsic limits* of science, which are the limits that belong to science i sofar as it is a human enterprise developed according to human features (la guage, structure, knowledge, method, activity, values, etc.). Thus, it may be th case that some limits of science can be overcome through scientific progres whereas other limits may be beyond scientific progress, due to the impossibi ity of the human being's reaching such aims.

Both aspects of the limits – contextual and intrinsic – can be considered i three different steps: science in general (i.e., features of any science), grou of sciences (the natural sciences, the social sciences, and the sciences of th artificial), and specific sciences (physics, chemistry, economics, sociolog pharmacology, computer sciences, etc.). The difficulties for the limits from on case to another can increase according to factors such as complexity (at lea epistemological and ontological),[9] which can be notable regarding some ph nomena (e.g., within the social sciences, in the case of economics, especial concerning prediction of phenomena in the long-run).

2 A Twofold Domain: Frontiers and Confines

Undoubtedly, the concern for frontiers or "barriers" and confines or "ceiling" i science is not new. At least, the distinction can be traced to its Kantian root where *Schranken* and *Grenzen* are considered in science, insofar as it is huma

8 Cf. Niiniluoto (1984a). On the concept of "scientific progress," see Gonzalez (1990).

9 Cf. Rescher (1998b, Ch. 1, pp. 1–24). These epistemological and ontological factors of comple ity can be seen in the sciences of the artificial, cf. Gonzalez (2012b, pp. 7–30).

knowledge.[10] In these roots we can find the need for the "positive" criteria regarding science (i.e., what science is or should be), which can also be enriched by the use of "negative" criteria (i.e., what science is not). The comparison with the negative criteria, which might be relevant for making clear what is impossible for scientific research, may be useful for "internal" purposes as well as for "external" ones.

2.1 *The Kantian Roots*

For Immanuel Kant, the twofold domain – *Schranken* and *Grenzen* – has a clear characterization regarding scientific knowledge. On the one hand, there is a secure path of doing science, which he wants to guarantee. Thus, its content should be distinguished from other kinds of human knowledge (such as metaphysics) [Kant 1787]. On the other, there are no limits concerning future knowledge, insofar as it is empirical knowledge, because the response given to a problem opens up a new question that should be responded to (Kant 1783).

Following the Kantian track, Karl Popper was also in this regard twofold in his philosophical approach, which develops several periods (Gonzalez 2004a, pp. 23–194; especially, pp. 41–65). Regarding frontiers or "barriers," he was interested in the demarcation problem.[11] In this central problem of his philosophy, his solutions involved a set of successive positions. Thus, he moved from the distinction between science and non-science to the difference between science and metaphysics, which included diverse conceptions of metaphysics (from a rather subjective view to a position in favor of objective grounds) [Gonzalez 2004a, pp. 23–194; especially, pp. 43, 45, 48, and 62]. Concerning confines or "ceiling" in science, Popper was more consistent in his appeal to science as an open endeavor. He uses "unended quest" as a motto for his intellectual trajectory (Popper 1976/1992).

Rescher also follows a Kantian track concerning the limits of science. (a) He recognizes a legitimate realm of human activities outside science. This includes problems that cannot be dealt with by scientific methods, because they are beyond its borders. Thus, aspects of human creativity, such as artistic expression (music, poetry, theatre, etc.), can communicate messages that cannot be articulated through scientific language (Rescher 1999a, p. 240). (b) He also defends the absence of confines for several reasons. Among them are two interrelated problems: the unfeasibility of scientific completeness, insofar as

10 These Kantian roots are considered within a Popperian approach in Radnitzky (1978). See also Radnitzky (1980).

11 On the historical background regarding the demarcation problem, see Nickles (2013).

each solution opens the door to a new question that also requires an answer, and the difficulties in achieving a perfect science, because he thinks that it is not possible to predict future science now (Rescher 1999b, Ch. 7, pp. 94–110). In addition, the complexity of the phenomena to be studied is another reason for leaving scientific research open to the future instead of closing of it.

2.2 *From the "Positive" Criteria to the "Negative" Criteria*

Nevertheless, looking at the Kantian roots, it should be emphasized that science is more than just human knowledge. Indeed, science includes a set of components,[12] of which knowledge is just one. Each component can be considered from the point of view of limits. The analysis of the components can give "positive" criteria of what science is or should be as well as "negative" criteria (i.e., what science is not now or cannot be any time in the future). These criteria require us to take into account that science is a human structure as well as a dynamic endeavor. Consequently, these components of a science, which are pointed out here, should be considered in both counts:

(i) Science typically possesses a specific language. Its terms have sense and reference that are generally precise and accurate. Thus, this language can be distinguished from other languages (artistic, philosophical, etc.), and its terms are limited: their sense might be modified over time and the reality referred to can change as well. (ii) Science is articulated in scientific theories, where there is a well patterned internal structure (at least in the most developed theories), which is nevertheless open to later changes. In this regard, logical analysis can be used to establish criteria for distinguishing these theories from other human constructions (such as cultural presentations) as well as criteria for the limits of the plausibility of these theories (as already done in formal sciences).

Furthermore, (iii) science is qualified knowledge. Its content has more rigor, in principle, than any other human knowledge. Scientific knowledge has so far been pivotal for the philosophical reflections on the frontiers or "barriers" and confines or "ceiling" in science. (iv) Science consists of a human activity that follows some methods (normally these processes are deductive, although many authors also accept inductive methods), and this appears as a dynamic activity (of a self-corrective kind). Methodological analysis can be used to distinguish "scientific processes" from "judgmental procedures" or uses of "phronesis" (Gonzalez 2015).

Besides these components of a science, which have been influential over a long period, there are other elements that have been emphasized in recent decades. (v) The reality of science comes from social action (which includes

12 The set of components of science pointed out here are in Gonzalez (2013b, pp. 15–16).

we-intentions).[13] It is an activity whose characteristics are different from other human activities in its assumptions, contents, and limits. Thus, philosophy of science should give the differences between scientific activity and other activities (such as artistic, philosophical, etc.). In addition, there are ontological limits to science, insofar as this science is *our* science (Rescher 1992).

(vi) Science has specific aims – where cognitive ones are particularly important – to guide, under the influence of values, its endeavor of research both in the formal sphere and in the empirical area. In this regard, axiology of scientific research should evaluate which ends are scientific and which are not. In addition, it should think of the ends that are achievable now and in the future (Gonzalez 2014a). (vii) Science can have ethical evaluations, insofar as the scientific enterprise is a free human activity, where certain values might be related to the process itself of research (honesty, originality, reliability, etc.) and some values can be connected with other activities of human life (social, cultural, economic, etc.). The concern for ethical limits of scientific research is clear regarding human beings and our society (Gonzalez 1999b).

From the point of view of the analysis of limits proposed here, each component of a science (language, structure, knowledge, method, activity, values, etc.) can be considered regarding the twofold domain (frontiers and confines). This can be followed by the study of the peculiarities of groups of sciences (natural, social, and artificial) as well as of the specific features of concrete sciences (physics, economics, computer sciences, etc.). The analysis can be made in terms of the structural viewpoint and in terms of the dynamic perspective of science. Consequently, this requires considering the contextual limits and the intrinsic limits.

3 The Discussion on the Frontiers or "Barriers" (*Schranken*)

Until now, the discussion on the frontiers or "barriers" of science has commonly moved in one of these directions. (I) There is no need for the criteria of distinguishing between science and non-science – and, especially, between science and philosophy –, for one of the following reasons: a) science can solve all the legitimate problems raised in academic terms (scientism); and b) the discussion is not worthy insofar as there is no actual point of comparison, because science is a part of a large academic project (as was the case with the classic Greeks or the Hegelian idealism, where philosophy embraced the scientific realm). (II) The discussion has had interest for a while, because it helped

13 See, in this regard, Tuomela (1991, 1996a, and 1996b).

to clarify what "science" is, but scientific progress and its interweaving with technology (including the specific proposal of "techno-science") means that the discussion no longer has sense, insofar as the present context is completely different from the past (for example, in comparison with Kantian times). (III) The discussion on the frontiers or barriers has some interest, but the criteria are not clear enough or cannot be established in universal terms (and, therefore, cannot have timeless character).

Yet, in spite of all these formulations of these objections – either structural or dynamic – we can think of criteria for the frontiers or barriers of science in its comparisons with other disciplines (philosophy, art, etc.). First, the existence of some kind of frontiers is something required by philosophers of science and scientists, who need to distinguish science form pseudoscience,[14] to make a difference between science and science-fiction (Barwick 2013, pp. 357–373; especially, pp. 357–360), etc. Second, the criteria are needed for a large number of organizations (public or private institutions) and professions that, in one way or another, accepted – at least implicitly – frontiers or barriers between science and non-science.

3.1 *The Need for Criteria for the Frontiers or "Barriers"*

Certainly, scientists need criteria to distinguish their activities from pseudoscientific undertakings and to make explicit possible academic frauds. Scientists are the people most interested in developing something that, from the theoretical and practical viewpoints, can have all the guarantees to be a "science." These criteria are assumed by them in their daily practice, in terms of explanation and prediction as well as in terms of prediction and prescription, and in the application of science. These criteria are also assumed by the referees that peer review scientific publications.

Obviously, philosophers of science need a concept of "science" and to be aware of the differences with other concepts in order to characterize the components of a science (such as language, structure, knowledge, activity, values, ...). This requires considering what science is but also what science ought to be. In addition, there are some other relevant features (such as objectivity, critical attitude, autonomy, and progress) [Niiniluoto 1984a, Ch. 1, pp. 4–7] that accompany robust scientific theories and reliable evidence. Thus, scientific progress is possible when scientists are seeking objectivity by means of the self-correctness of scientific procedures.[15] This allows us, in principle, to get

14 See, for example, Pigliucci and Boudry (2013). On the characteristics of a scientific paper see Suppe (1998a and 1998b), Lipton (1998), and Franklin and Howson (1998).

15 On self-correctness of scientific procedures, Rescher has insisted. See, for example Rescher (1978).

truthlikeness in scientific knowledge and to be able to give a solution to concrete problems in the practical sphere.

Subsequent to the recognition by scientists and philosophers of the need for some criteria regarding the possible frontiers or barriers of science, some proposals (such as Larry Laudan's demise of the demarcation problem, now and for the future)[16] can be seen as a philosophical mistake. On the one hand, this is a relevant philosophical problem as such: it deals with a key issue – at least for scientists and philosophers – of the notion of "science" and its boundaries; and, on the other, it has many practical consequences that should not be dismissed or ignored on purpose.

Also the criteria of a demarcation between science and non-science are assumed – in one way or another – by organizations (such as public or private institutions) and by professionals of diverse fields. In the first group, there are the referees that make the evaluation of research projects, the entities that endorse the research projects, the publishing houses that print books and journals devoted to scientific issues, the cultural societies and academic institutions that organize lectures and conferences about scientific topics, libraries, etc. In the second group are sociologists of science (in order to ascertain when a community is scientific instead of non-scientific), psychologists of science (when they try to figure out the processes of the scientists when they make a "scientific discovery"), historians of science, economists of science, etc. Therefore, the problem of demarcation underlies the daily work of an important number of professions.

Concerning the demarcation problem, Niiniluoto has defended that distinction between science and non-science is primarily descriptive, whereas the difference between science and pseudoscience has a prescriptive orientation: "the demarcation between science and *non*-science is primarily a matter of conceptual clarity and does not have an evaluative function. There are numerous human activities which are not – and do not pretend to be – 'scientific' (…) but still are valuable by their own standards. However, the demarcation between science and pseudoscience has the normative function of separating scientific and *un*-scientific activities from each other" (Niiniluoto 1984a, p. 2).

3.2 *Criteria According to Components of a Science*

Using the criteria based on the components of a science pointed out (language, structure, knowledge, method, activity, values, etc.), we can overcome some

16 Laudan claims that "there is no demarcation line between science and non-science, or between science and pseudo-science, which would win assent from a majority of philosophers. Nor is there one which *should* win acceptance from philosophers or anyone else," Laudan (1983, p. 112).

attempts of diluting or denying the need for the frontiers or barriers of science, such as Laudan's proposal to dismiss the problem of demarcation (Laudan 1983) or certain naturalist conceptions, mainly epistemological and methodological, in favor of establishing a complete continuity between science and philosophy.[17] These options move between a defense of a different level of generality between science and philosophy, without any constitutive difference as general intellectual project, and a clear-cut support of scientism, i.e., the assumption that science has the solution to any relevant intellectual problem, which makes philosophy unnecessary as a serious academic undertaking.

Semantically, we can distinguish a referent that is real (i.e., it actually exists, it has been in the past or it maybe so in the future) from a fictional or non-existent object. *Fiction* is by definition something that is not real, and its reference cannot be actual in any relevant sense (i.e., "entities" such as Pegasus). Thus, "science-fiction" is by definition "non-science," even though it is clear that scientific creativity requires the use of imagination. Meanwhile, idealization belongs to a different conceptual facet than fiction,[18] insofar as it is assumed that scientific models (mainly, the prescriptive models) have some degree of idealization, at least in the sense of simplification of the reality.

Although logic does not have now the relevance for philosophy of science that it used to have for several decades of the twentieth century,[19] it is still useful for the analysis of the limits of science. Logic can establish criteria for something that is conceptually impossible and, therefore, it follows that something that is beyond the range of the scientific enterprise is logically impossible. In this regard, logic can contribute to establish what can be considered as outside of the scientific realm.

Moreover, using some logical criteria, philosophy of science can think of the processes of research in terms of limits. On the one hand, there is the issue of the possible limits of deductivism in order to develop science (Grünbaum and Salmon, 1988). This has been a topic of discussion since at least the

17 The proposals of "scientific philosophy" are not new. What varies is the dominant scope used to be physical, then move to biological, and nowadays is mostly neurological.

18 Ann-Sophie Barwich mantains that "fiction (...) refers to the role played by particular methods of model building such as abstractions, idealisations and the employment of highly hypothetical entities," Barwick (2013, pp. 357–358). It seems to me that, in the context of models of science, which may be descriptive and prescriptive, "abstraction" and "idealization" have a useful role. Meanwhile, "fiction" is different from them, both in terms of sense and reference.

19 The logico-methodological approaches were dominant with the verificationist and falsificationist conceptions, which were very influential from mid-twenties to mid-sixties. See Gonzalez (2010, Part I, pp. 19–89).

mid-nineteenth century (Mansel 1853). On the other hand, there is the analysis of induction in terms of its reliability for science, and, therefore, as a limit for scientific progress. Although induction is not always presented in a logical setting,[20] it is clear that the logical criteria have been used in the discussions on induction (and on probability). This has been the case especially for the reliability of inductivism, in order to have science according to the present standards.[21]

Epistemologically, it should be possible to distinguish reliable knowledge regarding an entity from non-reliable knowledge or a presumptive knowledge about a non-entity. Scientific knowledge is, in principle, a content that goes beyond the standards of ordinary knowledge. Furthermore, it has or can have a guarantee regarding its objectivity, because this knowledge – in empirical sciences – is related to proprieties of what is real. In principle, scientific knowledge is neither purely subjective nor merely intersubjective knowledge.

Methodologically, scientific processes should have tests regarding their reliability. Thus, the processes should be involved in cases of repeatability, which in empirical sciences (natural, social, or artificial) are either in terms of observation or experimentation. These processes should enlarge human knowledge a rigorous way, either through explanation and prediction or by means of prediction and prescription. Consequently, it should be possible to distinguish "procedures" regarding prediction that are merely judgmental or based on subjective estimations from "methods" that have processes developed according to well-established patterns (Rescher 1998a, pp. 85–112. See especially Gonzalez 2015, Ch. 10, pp. 255–273).

Axiologically, there is now consensus that science is not "value-free" (Gonzalez 2013c). Moreover, science actually has – and also should have – values that are characteristic of this human enterprise. Primarily, these values are related to "internal" contents, mostly in the cognitive domain (such as truth, reliability, consistency, etc.), and, secondarily, these values are connected to "external" aspects (social, cultural, economic, ecological, etc.). These values are commonly assumed by the scientists in their daily practice. Some of them have support in human needs, whereas other values are built up following human options and preferences (Rescher 1999a, Ch. 3, especially, pp. 90–93).

---

20 This is the case at least since the controversy between William Whewell and John Stuart Mill on induction.

21 The discussions on the problem of induction have a very long tradition, and were very strong after Popper's attempts to dismiss induction as a scientific method based on some logical criteria.

Ethically, science is under scrutiny insofar as it is developed upon free human actions (Gonzalez 1999a). Ordinarily, science is made following human decisions oriented to some ends and according to some means that, in principle, are freely chosen. Thus, there are two lines for the barriers or frontiers in ethical terms: endogenous and exogenous. First, endogenous limits are those based on the scientific activity itself (i.e., as a human free endeavor, which should lead to honesty regarding the data obtained);[22] and, second, exogenous limits are those that come from "outside" of scientific activity (social, cultural, etc.). On the one hand, each scientist should be aware of what, while researching, he or she cannot do according to ethical principles; and, on the other society – through covenants or regulations – can establish what, concerning scientific research, is off-limits.[23]

Even though the set of criteria (semantic, logical, epistemological, methodological, ontological, axiological, and ethical) proposed here for the frontiers or "barriers" of science are richer than previous criteria (mostly focused on epistemological, methodological, and ethical aspects), there are still clear difficulties in establishing a neat line to distinguish what is science and what is not. The concern is related to the structural viewpoint and the dynamic perspective, because both are needed for this issue. In addition, the criteria used to distinguish the frontiers or "barriers" should not be merely descriptive: they should have a prescriptive feature as well.

Regarding the criteria – and taking into account the existence of scientific progress – there are at least three possibilities at stake. (a) It may be easy to separate what is pseudoscience from what is science (such as astrology and astronomy), but it is more difficult to establish the edge between proto-science and actual science.[24] (b) Some topics that initially were philosophical can later on become scientific (as has been the case with cognitive sciences). (c) There are sometimes overlapping of a scientific approach and a philosophical conception (in physics, psychology, economics, etc.), which could be for years. Thus, following the dynamic engaged by the scientific progress, the simple ide

22 Periodically, there is news on research misconduct both in basic science and in applie science. A recent example is what Haruko Obokata (RIKEN Center for Development Bio ogy in Kobe, Japan) claimed regarding being able to create pluripotent cells by exposin ordinary, non-stem cells to weak acids, physical squeezing and some bacterial toxins. *Th Economist*, v. 411, n. 8891, (June 14th–20th, 2014), pp. 70–71.

23 An analysis of many aspects of the ethics of science is in Agazzi (1992) and Gonzale (1999b).

24 Besides the philosophico-methodological concern, there is also an institutional aspec which has many practical consequences: When is a new subject ready for school, at th university (or even for a full academic course in a degree or master)?

of a neat line to draw the limits should be replaced by a more complex philosophical characterization of this problem.

Recent developments of interdisciplinary, multidisciplinary, and transdisciplinary studies can involve additional difficulties for the problem of "barriers" or frontiers.[25] In addition, the analysis needs to recognize the existence of constant links between philosophy, science, and technology. But this recognition of interrelation should not be an obstacle to conceptually distinguishing them. Consequently, the interrelation may be compatible with a characterization of what is science and what is not science, looking at this issue from a structural viewpoint and a dynamic perspective.

4 The Frequent Interest in the Confines or "Ceiling" (*Grenzen*)

To some extent, there is a decline in the interest for the frontiers or barriers of science in recent decades while there is still a lot of attention to the confines or ceiling of science in several directions. First, there is a straight forward reflection on this topic through the idea of an impossibility of a "perfect science." This leaves an always open door to the future, insofar as there cannot be a "complete" science.[26] Second, there is an analysis of the difficulties to establish "confines" of science due to the problems caused by complexity, both epistemological and ontological, which may be structural and dynamic.[27] Third, there is still available the Popperian idea that we do not know what we will know in the future, which reappears frequently in the discussion on scientific prediction and its limits, mainly in the social sciences.[28]

Kantian roots are still solid in this regard. Thus, there is frequently support for the idea of the absence of confines or ceiling of scientific research. This is the case at least in the sense of openness of scientific knowledge to the future and of recognition that we are not aware now of possible new referents for scientific research. Consequently, we can make meta-predictions about the reliability of scientific predictions regarding phenomena of the short-, middle- and

25 On transdisciplinarity, see Madni (2007).

26 This is the line commonly followed by Nicholas Rescher in his books on *The Limits of Science*.

27 On the structural complexity, see Gonzalez (2011a). On the dynamic complexity, see Gonzalez (2013a).

28 Popper was concerned about this topic while he wrote *The Poverty of Historicism* (1957), which open a very long line of discussions on the possibility and limits of scientific prediction in the case of the social sciences.

long-run. Certainly, there are "novel facts" in the ontological, epistemological, and heuristic ways. These possibilities of *novel facts* are particularly relevant for sciences such as economics or communication sciences (Gonzalez 2014b), which are social sciences as well as sciences of the artificial.[29]

Again, as was the case in the other limits of science (barriers or frontiers), the proposal regarding the limits that science might have from the viewpoint of confines or "ceiling" can be made in "internal" terms or in "external" terms. In the first case, the existence of boundaries of scientific research is assumed due to its "internal" processes (mainly of epistemological and methodological kinds). Thus, we can find a final point of scientific research (i.e., something beyond which is not possible to go). To some extent, this was the conception of the Vienna Circle, insofar as "verification" was accepted as a final end of research. In the second case, there are "external" decisions – social, cultural, political, economic, etc. – that can involve confines. This was the assumption made by sociologists in favor of the finalization in science (*Finalisierung der Wissenschaft*).[30]

4.1 *Limits of Science and the Unavailability of a Perfect Science*

The idea of an unavailability of a perfect science can be used in two different directions. On the one hand, the impossibility of reaching a "perfect science" can be understood in the sense that there is a ceiling in this human knowledge and the scientific methods. Thus, we cannot go further with science because human knowledge and processes are bounded and, therefore, they are *eo ipso* limited now and in the future. On the other hand, the impossibility of getting a "perfect science" can be used for just the opposite option. It means that we cannot get a ceiling or any kind of confines, insofar as any scientific knowledge and the processes of research can be revised. Consequently, we do not know what the future science can offer us (mainly in terms of problems, methods, and results).

It happens that, periodically, the first approach – understood as a proposal of internal limits of science – is launched. From time to time, there are concerns over the end of scientific discoveries. Some years ago, Steve Durlauf pointed out that "a recent spate of writings has argued that in a fundamental sense, science is reaching limits" (Durlauf 1997, p. 31). He is referring to John Horgan, who launched a broad attack regarding the possibility of future fundamental breakthroughs (Horgan 1996). Other authors include John Ziman, who thought that funding limitations will prevent the development of sufficiently

29 On the case of economics, see Gonzalez (2008).

30 This was the Starnberg group: cf. Böhme, Daele, and Krohn (1973/1976).

strong empirical evidence to provide the stimulus to important advances (Ziman 1996).

In addition, Colin McGinn's conception leads one to think that scientific investigations as well as philosophical reflections have limits in the study of consciousness (McGinn 1993). In this context, there is an emphasis on phenomena that some consider that may be insuperably difficult for humans to explain or predict as well as to predict and prescribe in scientific terms. Commonly, a large group of sciences (mainly in the sphere of the social sciences and, occasionally, in the realm of the sciences of the artificial) are severely criticized for their results: they are not considered as "fully" scientific.

Durlauf considers that the central claims of this line of thought, which is in favor of limits as confines, are twofold. (1) It is argued that for many areas of scientific research (notably physics, chemistry, and biology) there is now a basically accurate view of nature. In this regard, the theory of evolution, the theory of relativity, and quantum mechanics are taken as primary examples of scientific success.[31] (2) It is argued that in those areas where scientific research cannot claim great empirical successes, fundamental advances are impossible. This is because of their theories, insofar as they do not admit testable empirical predictions, or because of the experiments that discriminate between theories, inasmuch as they cannot be carried out (Durlauf 1997, p. 31).

However, Durlauf argues that the twofold line of thought in favor of limits of science is clearly unpersuasive. (i) He thinks that there is no basis for concluding that a successful science is at the end of major breakthroughs. (ii) He rejects the claim that social sciences, as examples of supposedly unsuccessful sciences, are incapable of producing successful models. As a professional economist, he considers that this frequent claim is unsupported. Moreover, he sees it more plausible to have more problems to identify bounds on science than that there are to reach such bounds. Consequently, he maintains that it is impossible to ever identify such possible limits of science (Durlauf 1997, p. 31).

Meanwhile the external limits of science in terms of confines were drawn by the sociologists of the Starnberg group (G. Böhme, W. van den Daele, W. Krohn, …). They thought that scientific boundaries can be established from a social standpoint, which might be of political kind.[32] This approach used a conception of the scientific development based on three stages, which connect

31 As a consequence of this viewpoint, further advances in explanation and prediction of physical, chemical, and biological phenomena will be derivative of these existing broad theories.

32 On this orientation on the limits of science, cf. Van den Daele, Krohn, and Weingart (1977); Krohn, Layton, and Weingart (1978); Böhme (1980); and Shäfer (1983).

to some ideas from Thomas Kuhn's initial methodological period, focused on the structure of scientific revolutions.[33]

These sociologists scheduled three main steps in their approach: (a) an exploratory phase or previous to the existence of a paradigm, (b) a period guided by a paradigm (or structural time), and (c) a later stage "post-paradigm" or time for extension of the applications. The external aims to the science were considered as the actual guide for the development of a scientific theory. This view has had a direct repercussion on how scientific progress is understood. In addition, it has had a clear-cut incidence in the issue of the autonomy of science (Gonzalez 1990, pp. 91–109; especially, pp. 100–104).[34]

Underneath this conception of the external limits of science there is a denial of the demarcation problem, because this sociologist approach to science does not conceive of a clear distinction between a scientific activity and other kinds of human activities. Consequently, the aims of science in the short-run and the ends of science in the long-run can be established outside the scientific community and upon sociological criteria. Even though they claimed to be using central Kuhnian ideas, they go far beyond Kuhn's views on the contextual aspects of scientific activity, because he recognizes the relevance of internal aspects of the scientific endeavor (including the importance of quantitative predictions) [Kuhn 1997/2000].

Rescher champions the defense of the absence of confines of science. He does this task regarding the internal limits (mainly, using epistemological criteria with a methodological repercussion). He uses the unavailability of the perfect science in the second option pointed out earlier. He thinks of four main conditions for a perfect science: (i) erotetic (question-answering) completeness or solving all the questions, (ii) predictive completeness or predicting accurately those eventuations that are predictable, (iii) pragmatic completeness or providing the means for whatever is feasible for human beings, and (iv) temporal finality or not having room for substantial changes (Rescher 2012, p. 154). He maintains that these conditions cannot be satisfied.

For Rescher, the key for thinking of a science without confines is epistemological. This openness to the future of scientific knowledge has methodological consequences and needs to consider ontological factors. In this regard, he assumes certain aspects. I) There is no possibility of a "complete" science, insofar as we do not know what the future of the science will be (Rescher 1999b, Ch. 7, pp. 94–110. See also his Ch. 1, pp. 5–18). II) A perfect science includes that every scientific problem will be solved, which is unthinkable if we look at

33 Regarding his views and the relation to this group, see Gonzalez (2004b).

34 On the relation to the applied science, cf. Niiniluoto (1984b, pp. 234–235).

the scientific progress (new problems, novel models, etc.) and we take into account what has happened in the history of science.[35] III) The interaction of the researcher with reality is not exhaustible, insofar as there is complexity and a technological escalation to deal with phenomena (mainly from nature, which is his main focus of attention) [Rescher 1999b, Ch. 4, pp. 43–65].

4.2 *Limits Due to Complexity*

If scientific research is seen as an interaction between the researcher – an individual or a group – and the reality (natural, social, or artificial) following some methods, then the existence of boundaries for scientific processes and the complexity of the real world can be considered as a source of the unviability of a universal methodology. This involves the existence of methodological limits to science because of the question of complexity, which affects problems, methods, and results of the scientific research (Gonzalez 2012a). Besides this internal perspective of the analysis of scientific processes, there is an external viewpoint: the issue of complexity requires adequate organizations (public or private institutions) that are able to deal with this increasing horizon of difficulties of this social undertaking.

Unquestionably, the interaction between the researcher and reality requires specific language in order to identify such complex systems (natural, social, or artificial) and their components and subcomponents within a given whole. In this regard, we can consider an idea from Ludwig Wittgenstein in his *Tractatus*, when he connects the limits of the language and the limits of the world.[36] We can think that our intellectual horizon given by science depends on our scientific language, and our scientific language involves the boundaries of the world given by the scientific words available to identify the referents.

On the one hand, there is the initial problem of getting the scientific terms with the sense and reference able to identify such complex systems (natural, social, or artificial) and, thereafter, be able to reidentify them as the same previously identified ones.[37] On the other hand, there is the additional difficulty of "saying" and "showing" in science, because sometimes science can be "saying" (for example, in prediction of novel facts in an ontological characterization) without being capable of "showing" something (e.g., any kind of proof or

35 He devotes a whole chapter – number 10 – to this issue, cf. Rescher (1999b, pp. 145–165).

36 The original sentence is focused on the individual: "*Die Grenzen meiner Sprache* bedeuten die Grenzen meiner Welt," Wittgenstein (1922, n. 5.6).

37 The idea of identification and reidentification of referents is developed by Peter Strawson in his "descriptive metaphysics" and analyzed in Gonzalez (1986).

evidence in favor of an statement). This situation can involve some contextua limits.

Logically, organized complexity – where may be an internal hierarchy o even a poly-hierarchy in the complex system – is not the same than a disorga-nized complexity (i.e., an event or occurrence that might involve phenomena such as anarchy, volatility or chaos), which poses serious obstacles to predict-ability (Rescher 1998a, pp. 134–135). The limits in terms of confines is not eas to be thought of for events when they have organized complexity. But it i harder still in the cases that involve disorganized complexity, because we nee to grasp the processes going of before we can give an explanation or prediction

Undoubtedly, the features of complexity can be available according to dif-ferent forms of scientific activity. Thus, they can be considered by taking int account the three levels of analysis indicated in this paper: science, in gen eral, a group of sciences, or a specific science. It seems that complexity can b considered, in principle, by any of the disciplines related to nature, social an artificial worlds.[38] Furthermore, complexity is initially twofold in this sphere it can be focused either from the structural perspective or from the dynami viewpoint,[39] which certainly involves the possibility of emergent properties.[4

Even though these features of complexity are mostly internal boundaries t the unviability of a universal methodology, they can also be thought of fron an external angle, insofar as science is a human activity in a social setting (his torical, cultural, economic, political, ecological, etc.). The study of a structura perspective of complexity is commonly made regarding the framework or con stitutive elements present in a group of sciences or in a specific science. Mean while, the analysis of dynamic viewpoint of complexity is related to chang over time of the motley elements involved in that collection of sciences or th specific science, taking into account the forces generating the change.[41] Bot aspects – structural and dynamic – are mainly internal to science, but they ar also interrelated with the factors due to the social milieu.

Obviously, both sides – structural complexity and dynamic complexity and their dependence on internal features as well as on external factors lead t

38 See, in this regard, Mainzer (2007).

39 On complexity from a *dynamic point of view*, see Gonzalez (2013a).

40 "The prospects for the emergence of an effective complex system are much greater if has a nearly-decomposable architecture," Simon (2001, p. 82).

41 I think that these categories of structural and dynamic can be used to articulate lists kinds of complexity such as "multilevel organization, multicomponent causal intera tions, plasticity in relation to context variation, and evolved contingency," Mitchell (200 p. 21).

the problem of the limits to science. On the one hand, they make it harder to have a final solution to a problem when complexity is involved (even though being a "complex system" is a matter of degree, by definition it is something that is not simple); and, on the other, the existence of complexity makes it more difficult to say in advance when scientific research can be considered as being in its confines or ceiling (it might be a contextual limit rather than an intrinsic limit).

Obstacles to the future science based on complexity are of different kinds, but the main ones are epistemological and ontological aspects. Nicholas Rescher has made a relevant presentation of these options. He thinks of the *epistemic modes* of complexity as divided into three groups, where it is possible to find a formulaic complexity: (1) descriptive complexity; (2) generative complexity; and (3) computational complexity. Meanwhile he sees the *ontological modes* of complexity as distributed in three main groups: (a) compositional complexity; (b) structural complexity (in a strict sense); and (c) functional complexity.[42]

Taking all these aspects together, it seems rather obvious that they entailed methodological limits to science. This makes it easier to think of a variety of methods according to objects (the aspects of reality) and problems (the focuses of attention) [Gonzalez 2012a, p. 164]. In addition, these obstacles due to complexity make a precise and accurate statement on the boundaries of the future science really difficult. Furthermore, scientific processes as well as the selection of aims of research – and, eventually, the results – can be connected to values, both in internal terms (cognitive, etc.) and in external terms (social, cultural, etc.). The insistence on science as value-laden, instead of the old appeal to value-free (Gonzalez 2013c), opens the door to new values related to the research of structural and dynamic sides of complexity.

Although the set of aspects of complexity mentioned here is long enough to be concerned about the possibility of thinking of any kind of confines of science, there is another important facet at stake: the issues regarding ethical components of science. They can be endogenous and exogenous to scientific activity as a free human undertaking in a social setting. This realm is one of the cases where the reflection on the limits of science is more acute, especially when the field of research involves bio-medicine. The future will depend very much on which kind of ethical orientation (Kantian, utilitarian, personalist, etc.) is assumed by scientists as well as the societies where research is done. Thus, although some limits to science have been imposed in certain countries

42 See, in this regard, Rescher (1998b, p. 9).

using laws approved by parliaments (mainly in bio-medical topics), thereafter new legislation has been passed in a different direction.

5 Coda: The Relevance of Novelty

Novelty can be used as a key feature to understand that complexity makes research difficult but it cannot be the reason to establish a final stage of science in advance. Novelty makes clear that the verificationists were not aware of the historicity of science, open to new discoveries and conceptual revolutions as well as important institutional changes regarding science. Thus, if we see complexity from the point of view of historicity of science, which is a trait that accompanies scientific progress, then the task of intrinsic limits of science seems to be a really tough job, especially if there is any attempt to give specific details of something intrinsically impossible for science at any time in the future.

From the angle of novelty, the recognition of boundaries in science, because it is a human activity with components according to human capacities (language, structure, knowledge, methods, etc.), is compatible with the proposal of leaving science open to the future. Scientific creativity and technological innovation can contribute to deal with complex systems (natural, social, and artificial). The increasing presence of new scientific domains (proteomic, genomics, etc.) and the improvement in other fields already available (artificial intelligence, neuroscience, etc.) supports novelty as a guarantee of openness to the future.

Philosophically, (1) science is not boundless, insofar as we can think of "barriers" or frontiers with other fields (such as philosophy, art, etc.). These limits, which should not be thought of something that is rigid or already given (either in structural terms or in dynamic terms), affect basic science (explanation and prediction), applied science (prediction and prescription), and application of science. (2) Science can be considered from the viewpoint of confines or "ceiling." But the configuration of science (language, structure, knowledge, method, activity, values, etc.) is without this kind of limits, insofar as it is not perfect or complete in any of these components, and the dynamics of science is open to the future.

Through this way of rethinking the limits of science, the contributions made up till now (mostly epistemological, methodological, and ethical) are taken into account. In addition, the proposal made here can enlarge the analysis through new considerations (mostly semantic, logical, ontological, and axiological)

Furthermore, the levels of analysis of science are also enlarged with more emphasis on applied science and the application of science as well as by means of the stages of science, groups of sciences, and specific sciences. There is always the clear notion that contextual limits of science are different from intrinsic limits of science. This is a consequence of the relevance of novelty in science.

Concerning how science of the future might be, there is a net difference if we think of the short-run or the long-run. This includes some variations regarding prediction of the future science. a) We know some present limits regarding the kind of scientific discoveries possible in the short-run, but it is harder to anticipate with rigor what *aims* science can achieve in the long-run. b) The characteristics of the possible *contents* in the discoveries of the short-run seems feasible, whereas the contents of science in the long-run are really difficult to predict, or even impossible for us today. c) The *results* that can follow from these contents obtained through the research in the short-run seem plausible, whereas the results in the long-run are difficult to foretell, or even impossible from our present situation. The reliability of these predictions will increase insofar as there is a clear dominion of the variables involved in what science is and should be at the levels pointed out: science in general, group of sciences (natural, social, artificial), and specific sciences (physics, economics, pharmacology, etc.).

Consequent to the emphasis on novelty regarding the future of science, there might be a prospective vision made in descriptive terms or developed in prescriptive ones. When the characterization is *descriptive*, then the future of science can be thought of as the contextual possible boundaries of what scientists as human beings think that they are able to do at a given time in the future. Meanwhile the approach from a *prescriptive* view can try to figure out what scientists should do in future science. This includes when they should stop their research (for example, when it is really harmful for persons or societies) as well as when they should address the research in a different way (for example, because it might be offensive for the shared values of a given society).

Along with this prospective vision, when the orientation is prescriptive, the analysis might envision the path "internally" to the scientific community or the study can be proposed "externally" (i.e., from the society or from its authorities). In order to figure out the prescriptive direction, there should be a conception of the application of that future science, which needs to accepted by the organizations (public and private) devoted to scientific research. Even with an eventual agreement of the application of this policy, either explicit or implicit, it is hard to believe that the line could be maintained in the long-run. Moreover, this application of that future science requires a clear idea of the difference between "bad science" and "good science,"[43] which is one step higher than the distinction between "science" and "non-science" and, obviously, is located above the difference between "science" and "pseudoscience" (or false science).

References

Agazzi, E. 1992. *Il bene, il male e la scienza: Le dimensioni etiche dell'impresa scientifico-tecnologica*. Milan: Rusconi.

Barwick, A.-S. 2013. "*Science and Fiction*: Analysing the Concept of Fiction in Science and its Limits," in *Journal of General Philosophy of Science*, 44, 357–373.

Böhme, G. 1980. "On the Possibility of 'Closed Theories'," in *Studies in History and Philosophy of Science*, 11, 163–172.

Böhme, G., Daele, W. Van den, and Krohn, W. 1973. "Finalisierung der Wissenschaft," in *Zeitschrift für Soziologie*, 2, 128–144. Translated into English as: 1976. "Finalization in Science," in *Social Science Information*, 15, 307–330.

Cattell, J.M. 1896. "The Limits of Science," in *Science*, New Series, 4 (94), 573.

Durlauf, S.N. 1997. "Limits to Science or Limits to Epistemology?," in *Complexity*, 2, (3), 31–37.

Franklin, A. and Howson, C. 1998. "Comment on 'The Structure of a Scientific Paper' by Frederick Suppe," in *Philosophy of Science*, 65, 411–416.

Gonzalez, W.J. 1986. *La Teoría de la Referencia. Strawson y la Filosofía Analítica*. Salamanca-Murcia: Ediciones Universidad de Salamanca and Publicaciones de la Universidad de Murcia.

Gonzalez, W.J. 1990. "Progreso científico, autonomía de la Ciencia y realismo," in *Arbor*, 135 (532), 91–109.

Gonzalez, W.J. 1998. "Prediction and Prescription in Economics: A Philosophical and Methodological Approach," in *Theoria*, 13 (32), 321–345.

Gonzalez, W.J. 1999a. "Ciencia y valores éticos: De la posibilidad de la Ética de la Ciencia al problema de la valoración ética de la Ciencia Básica," in *Arbor*, 162 (638), 139–171.

Gonzalez, W.J. (ed.).1999b. *Ciencia y valores éticos*, monographic issue of *Arbor*, 162 (638).

Gonzalez, W.J. 2004a. "La evolución del Pensamiento de Popper," in Gonzalez, W.J. ed. *Karl Popper: Revisión de su legado*, (pp. 23–194). Madrid: Unión Editorial.

Gonzalez, W.J. 2004b. "Las revoluciones científicas y la evolución de Thomas S. Kuhn," in Gonzalez, W.J. ed. *Análisis de Thomas Kuhn: Las revoluciones científicas* (pp. 15–103). Madrid: Trotta.

43 On this distinction see the remarks in Nickles (2013).

Gonzalez, W.J. 2008. “Rationality and Prediction in the Sciences of the Artificial: Economics as a Design Science,” in Galavotti, M.C., Scazzieri, R. and Suppes, P. eds. *Reasoning, Rationality, and Probability* (pp. 165–186). Stanford: CSLI Publications.

Gonzalez, W.J. 2010. *La predicción científica: Concepciones filosófico-metodológicas desde H. Reichenbach a N. Rescher*. Barcelona: Montesinos.

Gonzalez, W.J. 2011a. “Complexity in Economics and Prediction: The Role of Parsimonious Factors,” in Dieks, D., Gonzalez, W.J., Hartman, S., Uebel, Th. and Weber, M. eds. *Explanation, Prediction, and Confirmation*, (pp. 319–330). Dordrecht: Springer.

Gonzalez, W.J. 2011b. “Conceptual Changes and Scientific Diversity: The Role of Historicity,” in Gonzalez, W.J. ed. *Conceptual Revolutions: From Cognitive Science to Medicine*, (pp. 39–62). A Coruña: Netbiblo.

Gonzalez, W.J. 2012a. “Methodological Universalism in Science and its Limits: Imperialism versus Complexity,” in Brzechczyn, K. and Paprzycka, K. eds., *Thinking about Provincialism in Thinking*, Poznan Studies in the Philosophy of the Sciences and the Humanities, vol. 100, Rodopi, Amsterdam/N. York,, pp. 155–175.

Gonzalez, W.J. 2012b. “Las Ciencias de Diseño en cuanto Ciencias de la Complejidad: Análisis de la Economía, Documentación y Comunicación,” in Gonzalez, W.J. ed. *Las Ciencias de la Complejidad: Vertiente dinámica de las Ciencias de Diseño y sobriedad de factores*, (pp. 7–30). A Coruña: Netbiblo.

Gonzalez, W.J. 2013a. “The Sciences of Design as Sciences of Complexity: The Dynamic Trait,” in Andersen, H., Dieks, D., Gonzalez, W.J., Uebel, Th., and Wheeler, G. eds. *New Challenges to Philosophy of Science*, (pp. 299–311). Dordrecht: Springer.

Gonzalez, W.J. 2013b. “The Roles of Scientific Creativity and Technological Innovation in the Context of Complexity of Science,” in Gonzalez, W.J. ed. *Creativity, Innovation, and Complexity in Science*, (pp. 11–40). A Coruña: Netbiblo.

Gonzalez, W.J. 2013c. “Value Ladenness and the Value-Free Ideal in Scientific Research,” in Lütge, Ch. ed. *Handbook of the Philosophical Foundations of Business Ethics*, (pp. 1503–1521). Dordrecht: Springer.

Gonzalez, W.J. 2014a. “On Representation and Models in Bas van Fraassen’s Approach,” in Gonzalez, W.J. ed. *Bas van Fraassen’s Approach to Representation and Models in Science*, (pp. 3–37). Dordrecht: Synthese Library, Springer.

Gonzalez, W.J. 2014b. “The Evolution of Lakatos’s Repercussion on the Methodology of Economics,” in *HOPOS: The Journal of the International Society for the History of Philosophy of Science*, 4 (1), 1–25.

Gonzalez, W.J. 2015. *Philosophico-Methodological Analysis of Prediction and its Role in Economics*. Dordrecht: Springer.

Grünbaum, A. and Salmon, W.C. eds. 1988. *The Limitations of Deductivism*. Berkeley: University of California Press.

Hammond, A.L. 1975. “Speaking of Science: Weisskopt and Limits of Science,” in *Science*, New Series, 188 (4189), 721.

Horgan, J. 1996. *The End of Science: Facing the Limits of Knowledge in the Twilight of the Scientific Age*. Reading, MA: Addison-Wesley.

Kant, I. 1783. *Prolegomena zu einer jeden künftigen Metaphysics, die als Wissenschaft wird auftreten können,* Riga: Johann Friedrich Hartknoch. Reprinted in Kant, I. 1991. *Gesammelte Schriften*, Berlin: Reimer, vol. 4, and Hamburgh: Reclam, 2001.

Kant, I. 1787. *Kritik der Reinen Vernunft*. Riga: Johann Friedrich Hartknoch, 2nd ed. Translated into English by Norman Kemp Smith. 1865. *Critique of Pure Reason*, N. York: St. Martin's Press.

Krohn, W., Layton, E. Jr., and Weingart, P. eds. 1978. *The Dynamics of Science and Technology*. Dordrecht: Reidel.

Kuhn, Th.S. 1997. "A Physicist who Became Historian for Philosophical Purposes," in *Neusis. Journal for the History and Philosophy of Science and Technology*, 6, 145–200. Reprinted as Kuhn, Th.S. 2000. "A Discussion with Thomas S. Kuhn," in Kuhn, Th.S *The Road Since* Structure: *Philosophical Essays, 1970–1993, with an Autobiographical Interview*, (pp. 255–323). Edited by James Conant and John Haugeland. Chicago: The University of Chicago Press.

Laudan, L. 1983. "The Demise of the Demarcation Problem," in Cohen, R. and Laudan L. eds. *Physics, Philosophy and Psychoanalysis*, (pp. 111–128). Dordrecht: Reidel.

Lipton, P. 1998. "The Best Explanation of a Scientific Paper," in *Philosophy of Science* 65, 406–410.

Madni, A.M. 2007. "Transdisciplinarity: Reaching Beyond Disciplines to Find Connections," in *Transactions of Society for Design and Process Science*, 11 (1), 1–11.

Mainzer, K. 2007. *Thinking in Complexity. The Computational Dynamics of Matter, Mind and Mankind.* Berlin: Springer, 5th edition.

Mansel, H.L. 1853. *The Limits of Demonstrative Sciences Considered in a Letter to the Rev William Whewell, D.D.*, William Graham, Oxford.

McGinn, C. 1993. *The Problems of Philosophy*. Oxford: Basil Blackwell.

Metcalfe, J.S. and Ramlogan, R. 2005. "Limits to the Economy of Knowledge and Knowledge of the Economy," in *Futures*, 37 (7), 655–674.

Mitchell, S.D. 2009. *Unsimple Truth: Science, Complexity, and Policy*. Chicago: The University of Chicago Press.

Nickles, Th. 2013. "The Problem of Demarcation: History and Future," in Pigliucci, M and Boudry, M. eds. *Philosophy of Pseudoscience: Reconsidering the Demarcation Problem*, (pp. 101–120). Chicago: The University of Chicago Press.

Niiniluoto, I. 1984a. *Is Science Progressive?* Dordrecht: Reidel.

Niiniluoto, I. 1984b. "Finalization, Applied Science, and Science Policy," in Niiniluoto, *Is Science Progressive?*, (pp. 234–235). Dordrecht: Reidel.

Pigliucci, M. and Boudry, M. eds. 2013. *Philosophy of Pseudoscience: Reconsidering the Demarcation Problem*. Chicago: The University of Chicago Press.

Planck, M. 1949. "The Meaning and Limits of Exact Science," in *Science*, New Series, 110 (2857), 319–327.

Popper, K.R. 1957. *The Poverty of Historicism*. London: Routledge.

Popper, K.R. 1976. *Unended Quest. An Intellectual Autobiography*. London: Fontana/Collins, (enlarged version London: Routledge, 1992).

Radnitzky, G. 1978. "The Boundaries of Science and Technology," in: *The Search for Absolute Values in a Changing World, Proceedings of the VIth International Conference on the Unity of Sciences*, vol. II, (pp. 1007–1036). N. York: International Cultural Foundation Press.

Radnitzky, G. 1980. "What Limits do Technology and Science Have?," in *Crítica. Revista Hispanoamericana de Filosofía*, 12 (35), 15–54.

Rescher, N. 1978. *Scientific Progress: A Philosophical Essay on the Economics of Research in Natural Science*. Oxford: Blackwell.

Rescher, N. 1984. *The Limits of Science*. Berkeley: University of California Press.

Rescher, N. 1992. "Our Science as *our* Science," in Rescher, N. *A System of Pragmatic Idealism. Vol. I: Human Knowledge in Idealistic Perspective*, (pp. 110–125). Princeton: Princeton University Press.

Rescher, N. 1998a. *Predicting the Future: An Introduction to the Theory of Forecasting*. Albany: State University of New York Press.

Rescher, N. 1998b. *Complexity: A Philosophical Overview*. New Brunswick, NJ: Transaction Publishers.

Rescher, N. 1999a. *Razón y valores en la Era científico-tecnológica*. Barcelona: Paidós.

Rescher, N. 1999b. *The Limits of Science*, revised edition. Pittsburgh: University of Pittsburgh Press.

Rescher, N. 2012. "The Problem of the Future Knowledge," *Mind and Society*, 11 (2), 149–163. DOI 10.1007/s11299-012-0099-8.

Shäfer, W. ed. 1983. *Finalization in Science: The Social Orientation of Scientific Progress*. Dordrecht: Reidel.

Simon, H.A. 2001. "Complex Systems: The Interplay of Organizations and Markets in Contemporary Society," in *Computational and Mathematical Organization Theory*, 7, 78–85.

Suppe, F. 1998a. "The Structure of a Scientific Paper," in *Philosophy of Science*, 65, 381–405.

Suppe, F. 1998b. "Reply to Commentators," in *Philosophy of Science*, 65, 417–424.

Tuomela, R. 1991. "The Social Dimension of Action Theory," in *Daimon*, 3, 145–158.

Tuomela, R. 1996a.*The Importance of Us*. Stanford: Stanford University Press.

Tuomela, R. 1996b. "Intenciones conjuntas y acuerdo," in Gonzalez, W.J. ed., *Acción e Historia. El objeto de la Historia y la Teoría de la Acción*, (pp. 277–291). A Coruña: Publicaciones Universidad de A Coruña.

Van den Daele, Krohn, W., and Weingart, P. 1977. "Political Direction of Scientific Development," in Mendelsohn, E. Weingart, P. and Whitley, R.D. eds. *The Social Production of Scientific Knowledge*. Dordrecht: Reidel.

Weisskopf, V.F. 1977. "Views: The Frontiers and Limits of Science: Modern Science is a Powerful Tool for Acquiring Deeper Insights into the World Around Us, But We Must Also Follow Other Avenues Toward Reality," in *American Scientist*, 65 (4), 405–411.

Wittgenstein, L. 1922. *Tractatus logico-philosophicus*, bilingual edition (German-English). London: Kegan Paul.

Ziman, J. 1996. *The Limits of Science*. Cambridge: Cambridge University Press.

Complementary Bibliography

Allen, P.M., Strathern, M., and Baldwin, J.S. 2007. "Complexity and the Limits to Learning," in *Journal of Evolutionary Economics*, 17, 401–431.

Almeder, R. 2008. "The Limits of Science, Realism, and Idealism," in Almeder, R. ed. *Rescher Studies. A Collection of Essays on the Philosophical Work of Nicholas Rescher*, (pp. 1–28). Heusenstamm: Ontos Verlag.

Anguelov, S. 1984. "Methodological Problems of the Social Sciences," in *Studies in Soviet Thought*, 27 (3), 263–265.

Babette, B.E., Bergoffen, D.B., and Glynn, S. eds. 1995. *Continental and Postmodern Perspectives in the Philosophy of Science*. Brookfield, VT: Avebury.

Barrow, J.D. 1998. *Impossibility: The Limits of Science and the Science of Limits*. N. York: Oxford University Press.

Bernard, L.L. 1929. "The Limits of the Social Sciences and their Determinants," in *The Journal of Philosophy*, 26 (16), 430–438.

Bonk, Th. 2008. *Underdetermination: An Essay on Evidence and the Limits of Natural Knowledge*. Dordrecht: Springer.

Brinkmann, K. 2011. *Idealism without Limits: Hegel and the Problem of Objectivity*. Dordrecht: Springer.

Carrier, M., Massey, G.J., and Ruetsche, L. eds. 2000. *Science at Century's End: Philosophical Questions on the Progress and Limits of Science*. Pittsburgh: University of Pittsburgh Press.

Casti, J.L. and Karlqvist, A. eds. 1996. *Boundaries and Barriers: On the Limits to Scientific Knowledge*. Reading, MA: Addison-Wesley.

Cattell, J.M. 1896. "The Limits of Science," in *Science*, New Series, 4 (94), p. 573.

Crowther, P. 2003. *Philosophy and Postmodernism: Civilized Values and the Scope of Knowledge*. N. York: Routledge.

DeCanio, S.J. 2014. *Limits of Economics and Social Knowledge*. N. York: Palgrave Macmillan.

Desroches, D. 2006. *Francis Bacon and the Limits of Scientific Knowledge*. London: Continuum.

Doppelt, G. 1990. "The Naturalist Conception of Methodological Standards in Science: A Critique," in *Philosophy of Science*, 57 (1), 1–19.

Etzioni, A. 2012. "The Limits of Knowledge," *Issues in Science and Technology*, 29 (1), 49–56.

Gauch, H.G. Jr. 2012. *Scientific Method in Brief*. Cambridge: Cambridge University Press.

Gleiser, M. 2014: *The Island of Knowledge: The Limits of Science and the Search for Meaning*. N. York: Basic Books.

Hendricks, V.F. 2001. *The Convergence of Scientific Knowledge: A View from the Limit*. Boston: Kluwer.

Hendry, J. 1950. "The Principle of Limits: With Special Reference to the Social Sciences," in *Philosophy of Science*, 17 (3), 247–253.

Jahnke, H.N. and Otte, M. eds. 1981. *Epistemological and Social Problems of the Sciences in the Early Nineteenth Century*. Dordrecht: Reidel.

Jasanoff, S. and Nelkin, D. 1981. "Science, Technology, and the Limits of Judicial Competence," in *Science*, New Series, 214 (4526), 1211–1215.

Karimi, K. and Hamilton, H.J. 2004. *Problem Solving with Limited Knowledge: Pruning Attributes in Classification Rules*. Regina: University of Regina.

Kirkman, R. 2002. *Skeptical Environmentalism: The Limits of Philosophy and Science*. Bloomington: Indiana University Press,.

Kraemer, J.D. and Gostin, L.O. 2012. "The Limits of Government Regulation of Science," in *Science*, New Series, 335 (6072), 1047–1049.

Lefevre, S.R. 1974. "Science and the Liberal Mind: The Methodological Recommendations of Karl Popper," in *Political Theory*, 2 (1), 94–107.

Little, D. 1993. "On the Scope and Limits of Generalizations in the Social Sciences," in *Synthese*, 97 (2), Empiricism in the Philosophy of Social Science, 183–207.

Marchetti, C. 1998. *Notes on the Limits to Knowledge Explored with Darwinian Logic*. Laxenburg: International Institute for Applied Systems Analysis.

Mattick, P. 1986. *Social Knowledge: An Essay on the Nature and Limits of Social Sciences*. Armonk: Sharpe.

Metcalfe, J.S. and Ramlogan, R. 2005. "Limits to the Economy of Knowledge and Knowledge of the Economy," in *Futures*, 37 (7), 655–674.

Mohan, Ch. 2000. *Frontiers of Expert Systems: Reasoning with Limited Knowledge*. Boston: Kluwer.

Muelder, W.G. 1979. "The Science of Limits and the Limits of Science," in *Selected Papers from the Annual Meeting (American Society of Christian Ethics)*, Twentieth Annual Meeting, pp. 1–22.

Munévar, G. 1981. *Radical Knowledge: A Philosophical Inquiry into the Nature and Limits of Science*. Indianapolis: Hackett.

O'Hara, P. 2010. *The Limits of Knoweldge*. Bloomington: Xlibris.

Ormerod, P. 2005. "Complexity and the Limits to Knowledge," in *Futures*, 7 (7), 721–728.

Pandit, G. 1991. *Methodological Variance: Essays in Epistemological Ontology and the Methodology of Science*. Dordrecht: Springer.

Paul, P.V. and Moores, D.F. eds. 2012. *Deaf Epistemologies: Multiple Perspectives on the Acquisition of Knowledge*. Washington, DC: Gallaudet University Press.

Perlman, J.S. 1995. *Science without Limits: Toward a Theory of Interaction between Nature and Knowledge*. Amherst, NY: Prometheus Books.

Przelecki, M. and Jadacki, J.J. eds. 2010. *Within and Beyond the Limits of Science: Logical Studies of Scientific and Philosophical Knowledge*. Warszawa: Naukowe Semper.

Roncallo-Dow, S., Uribe-Jongbloed, E., and Calderón-Reyes, I. 2013. "Research in Communication: Knowledge Limits and Limitations," in *Co-herencia*, 10 (18), 161–187.

Rosenberg, A. 1993. "Scientific Innovation and the Limits of Social Scientific Prediction," in *Synthese*, 97, 161–182.

Sanford, A.J. ed. 2003. *The Nature and Limits of Human Understanding*. N. York: T & T Clark.

Schlesinger, G.N. 1986. "On the Limits of Science," in *Analysis*, 46 (1), 24–26.

Spirito, U. and Heath, P. eds. 1952. "The Limits of Science," in *The Philosophical Quarterly*, 2 (8), 208–217.

Stanley, T.D. 2000. *Challenging Time Series: Limits to Knowledge, Inertia, and Caprice*. Cheltenham: Edward Elgar.

Starbuck, W.H., Holloway, S., Whalen, P.S., and Tilleman, S.G. eds. 2008. *Organizational Learning and Knowledge Management*. Northampton, MA: Edward Elgar.

Stelmach, J., Brozek, B., and Kurek, L. 2013. *Philosophy in Neuroscience*. Kraków: Copernicus Center Press.

Stokstad, E. 2008, "New Rules on Saving Wetlands Push the Limits of the Science," in *Science*, New Series, 320 (5873), 162–163.

Vitek, W. and Jackson, W. eds. 2008. *The Virtues of Ignorance: Complexity, Sustainability, and the Limits of Knowledge*. Lexington: University Press of Kentucky.

Weigel, G. and Madden, A.G. 1961. *Knowledge, its Values and Limits*. Englewood Cliffs NJ: Prentice-Hall.

Williamson, T. 2000. *Knowledge and its Limits*. Oxford: Oxford University Press.

Yaneva, D. 2006. "What Is Science? Methodological Pitfalls Underlying the Empirical Exploration of Scientific Knowledge," in *Journal for General Philosophy of Science, Zeitschrift für allgemeine Wissenschaftstheorie*, 37 (2), 333–353.

The Uncertain Frontier between Scientific Theories and Metaphysical Research Programmes

Juan Arana

Abstract

Until now, the problem of demarcation between science and philosophy has been one of the most fruitful issues regarding epistemology. At the same time, it has been also sterile, insofar as no reliable and widely accepted criterion has been established in this regard. Thus, we do not have clear boundaries which circumscribe scientific rationality and differentiate it from all others forms of reason. The discussion concerning the "limits of science" has one of the most influential views in this realm in the Popperian approach. There is here an evaluation of Popper's frontier between scientific theories and metaphysical research programmes. This opens the door to the final remarks on the uncertainty of the frontier.

Keywords

Frontier – demarcation – scientific theories – metaphysical research programmes – Popper

1 The Problem of Demarcation between Science and Philosophy

The problem of demarcation between science and philosophy – or, if one prefers, between science and metaphysics – has been one of the most fruitful and, at the same time, sterile issues studied by epistemology. Sterile, because it must be admitted that no reliable and widely accepted criterion has been established for marking the boundaries that circumscribe scientific rationality and differentiate it from all others forms of reason. Fruitful, because it has given rise to an infinite number of discussions, proposals, and polemics; in short, it has played the role of a stimulus that has triggered many other studies.

As a result, we are faced with a landscape characterized by lights and shadows. The parts most in shadow make us think of a continuous weaving and unraveling, of the necessity to reopen time and again questions that we had thought were definitively resolved. Dogmatism and skepticism are the two poles that the critical spirit oscillates between, and it is disheartening to

 | DOI 10.1163/9789004325401_003

discover that there is no way to rest at one or the other extreme, nor has it been possible to establish a stable equilibrium between them.

In his 1970 article *Falsification and the Methodology of Scientific Research Programmes*, later included in his book *The Methodology of Scientific Research Programmes,* Imre Lakatos summed up the results of this seemingly endless to and for: "just as some earlier ex-justificationists led the wave of sceptical irrationalism, so now some ex-falsificationists lead the new wave of sceptical irrationalism and anarchism" (Lakatos 1978, p. 91). An implacable destiny makes all epistemological proposals become complex and tangled until they lose all verisimilitude, as happened with falsificationism.

Here Lakatos distinguished between a dogmatic version, another which is naive, and a third which is sophisticated. Using these nuances he distinguishes up to three Poppers, the first of which "never published a word" (Lakatos 1978, p. 93). The natural tendency clearly points towards an increasingly jumbled state, with the result that there end up being not just as many opinions as there are heads that think, but rather plenty more. One understands and almost shares the opinion expressed by Roger Penrose that "philosophers tend to busy themselves with their own internal controversies" (Penrose 1996, p. 229).

2 The Discussion Concerning the "Limits of Science": The Popperian Approach

Since I do not forswear my status as a philosopher, I assume that in one way or another I will end up falling into the same vice. However, I will attempt to postpone this as much as possible. This article is located within a discussion concerning the "limits of science". Both the positivism of the 19th century and the neopositivism of the 20th century were based on the idea that science monopolizes the substantive knowledge provided by reason, since for these schools philosophy could, at best, exercise a critical-reflexive function.

Even in this latter case the situation remained conflictive, since science itself contains a self-critical dimension, and as a result it is not at all clear what the use is of a reflection that abstracts from content. In the category of formal sciences we already have logic, and perhaps mathematics. There would be, according to the supposition in question, sciences that would deal with the pure forms of thought, on the one hand, and positive sciences, on the other, with the latter's access to reality verified by way of experience. Philosophy would be relegated to the status of a residue of the past, when scholars believed in the legitimacy of "other" ways of accessing truth, or else as an area of interdisciplinary discussion, appropriate to the degree that science has not yet attained a definitive unification.

The postulate of the unity of knowledge is an ideal that the majority consider to be good and desirable, but very few have attempted to follow it through to its ultimate consequences, at least from the time of the death of Christian Wolff until now. What is typical has been to make a decisive split between the world of representations and the real world (or even the real world itself), locating them in separate spheres, in accordance with the Kantian model (or some other comparable model). The demarcationist strategies tend to leave aside this issue, which is typically philosophical, and attempt to map out the territory of science, where one supposes that reason can work in enjoyment of the fullness of its prerogatives. Everything else is tossed out into the "outer darkness", where one can stumble upon the most useless of the creations of the human mind: from metaphysics to magic, by way of palmistry and the study of flying saucers.

In this context Karl Popper represents an exception, since from the beginning he stated that his demarcation was not aggressive, and was far from attempting to treat everything that is not science as being equally ridiculous. "As it occurred to me first, the problem of demarcation was not the problem of demarcating science from metaphysics but rather the problem of demarcating science from pseudoscience. At the time I was not at all interested in metaphysics. It was only later that I extended my 'criterion of demarcation' to metaphysics" (Popper 1992, p. 41).

He has repeated several times that his intention was to objectivize the difference between science and the doctrines of the "false prophets" Marx and Freud, since certain personal experiences led him to question the solvency of the Marxist theory of history and of psychoanalysis (Popper and Kreuzer 1982, pp. 8–12). What he desired, therefore, was not a criterion for recognizing *what is scientific,* but rather only *what is rational.* Falsifiability, whether it be the version that Lakatos calls dogmatic, or else the methodological version, allowed him to make this first demarcation. Later he had to refine his position (Popper 1963 and 1979), because he needed to specify the scientific amongst other forms of rationality (specifically, metaphysics).

At that point all the nuances mentioned above came into play. In addition, however, the original criterion had to be reformulated, since if strict falsifiability cannot identify the limits of positive science, neither does it allow for sufficiently clarifying the borders that separate metascientific rationality from pseudosciences. In a way, Popper took a step backward, since the strength of his earliest proposal was diluted into the more generic and ambiguous notion of that which is susceptible to rational discussion:

"But if contradictions need not be avoided, then any criticism and any discussion becomes impossible since criticism always consists in pointing out contradictions either within the theory to be criticized, or between it and some facts of experience. The situation with psycho-analysis is similar: the psycho-analyst can always explain away any objections by showing that they are due to the repressions of the critic. And the philosophers of meaning, again, need only point out that what their opponents hold is meaningless, which will always be true, since 'meaninglessness' can be so defined that any discussion about it is by definition without meaning. Marxists, in a like manner, are accustomed to explain the disagreement of an opponent by his class bias, and the sociologists of knowledge by his total ideology. Such methods are both easy to handle and good fun for those who handle them. But they clearly destroy the basis of rational discussion, and they must lead, ultimately, to anti-rationalism and mysticism" (Popper 1971, II, pp. 215–216).

As he recalls in his intellectual autobiography, previously "I had not then realized that a metaphysical position, though not testable, might be rationally critizable or arguable" (Popper 1992, p. 150).

Employing Leibnizian terminology, it could be said that Popper defends a monadic conception of reality and seeks to close any other window of communication with the contaminated external environment. But science does not exist only within this sphere of meaning. And this is because science itself is an open reality, in progress, whose external territory unceasingly grows larger. At times it also passes through stages of contraction. Science has a temporal character, as does everything human: it is not a Parmenidean entity, but rather it fluctuates between before and after, between what was not and what still is not. Where were the theories of gravitation prior to Newton and Einstein? Where are the theories that will come into existence over the course of the 21st century and those that will be developed even later?

These ideas appear to collide with the peculiar Platonism of Popper and his *World 3*, but it should be kept in mind that within science a dimension of pure objectivity coexists with the fact that it is a living construction in which *World 1* and *World 3* interact, by way of the psychological instantiation that is represented by *World 2*. In accordance with the mature position of Popper, science and pseudoscience are, or should be, separated by an impassable barrier while between metaphysics and science there is a permeable membrane that permits osmosis to do its work. This is precisely the context where the concept of a "metaphysical research programme" plays an important role:[1]

"In science, problem situations are the result, as a rule, of three factors. One is the discovery of an inconsistency within the ruling theory. A second is the discovery of an inconsistency between theory and experiment – the experimental

1 According to Popper, he used this term in his lectures starting in 1949, and it first appeared in print in 1958. The "scientific research programmes" would thus constitute an attenuated version of the Popperian concept. See Popper (1992, p. 231, note 242).

falsification of the theory. The third, and perhaps the most important one, is the relation between the theory and what may be called the 'metaphysical research programme'. [...] In using this term I wish to draw attention to the fact that in almost every phase of the development of science we are under the sway of metaphysical, that is, untestable ideas; ideas which not only determine what problems of explanation we shall choose to attack, but also what kinds of answers we shall consider as fitting or satisfactory or acceptable, and as improvements of, or advances on, earlier answers. [...] I call these research programmes 'metaphysical' also because they result from general views of the structure of the world and, at the same time, from general views of the problem situation in physical cosmology. I call them 'research programmes' because they incorporate, together with a view of what the most pressing problems are, a general idea of what a satisfactory solution of these problems would look like [...]. They may be described as speculative physics, or perhaps as speculative anticipations of testable physical theories" (Popper 1982b, pp. 161–162).[2]

Popper provides a list of programs relating to physics. These include the block-universe of Parmenides, atomism, the geometrizing cosmology of Plato, Eudoxus and Euclid, essentialism and potentialism, Renaissance physics (Copernicus, Bruno, Kepler, Galileo), the model of the world as a watch (Hobbes, Descartes, Boyle), the dynamism of Newton and Leibniz, the fields of forces (Faraday, Maxwell), unified field theory (Riemann, Einstein, Schrödinger) and the statistical interpretation of quantum theory (Born) (Popper 1982b, pp. 162–164).

Elsewhere he proposes the idea that Darwinism also constitutes a metaphysical research programme (Popper 1992, pp. 167–180). It is noteworthy, in this regard, that the most relevant testable empirical prediction of the theory of natural selection, that of *gradualness,* was and is its most controversial point, ever since Kelvin's arguments about the age of the Earth, passing through mutationism, up to the theory of punctuated equilibrium or the neutralist theory of Kimura (Arana 2012, pp. 239–244).

3 An Evaluation of the Popperian Approach

A first evaluation of Popper's proposal would bear in mind that it proposes a non-exclusivist demarcationism, given that he does not seek to identify rational knowledge with science: what he is seeking to do is separate, with the

2 See also Popper (1983, pp. 131–46).

greatest accuracy possible, science from pseudo-science. In its first version this demarcation is a partial solution, even incomplete: it does not serve to clearly distinguish science from other legitimate manners of exercising reason. Therefore it is necessary to provide a second criterion of demarcation, or else to make more precise the first criterion in a more restrictive way. How? Coherence and empirical contrastability serve to delimit metaphysics and/or metaphysical research programmes, while what Lakatos calls *methodological falsificationism* (whether naive or sophisticated) filtered the theoretical *corpus* of reason in order to make objective that which properly and legitimately makes up the body of science.

This second possibility has the attractive feature of preserving the ideal of the unity of knowledge – albeit in a lax way – since between metaphysics, etc. and science there would be an analogous relationship to that which exists between natural language and a specialized language. The disadvantage – or, depending on the point of view adopted, the virtue – of this interpretation is that any other form of rational, non-scientific knowledge is transformed into a propaedeutic to science. If in times past philosophy was seen as the *ancilla theologiae,* today it is considered to be an *ancilla scientiae.* The novelty, in comparison with the various positivisms, is that the aid provided to science is no longer in the formal or reflexive terrain, but rather in the material and substantive, after the fashion of a heuristic and exploratory complement. Like the vanguard of an army, philosophy enters into unexplored regions, taking risks that often lead to failure, although its disasters constitute a condition of possibility for the successes later achieved by science.

Nevertheless, there is an alternative: the subsidiary role that metaphysics can play with regard to science is compatible with its own autonomy when it comes time to develop cognitive projects that would be ordered to other ends and would not need to be accredited in exchange for the aid that it could provide to the invention and development of scientific theories. This goes not only for metaphysics in general, but also for the metaphysical research programmes even though in these latter the heuristic functions are always necessarily present. For the scientist, the most interesting aspect of these programmes is instrumental, while for those who are not directly involved in research, their aptitude for managing interdisciplinarity has as much or more importance. Thanks to these programmes, metaphysical theories have the right to claim meaning and even truth, even when it is accepted that, in accordance with Popper's characterization, they are unable to be tested. In the end, their heuristic capacity provides them with an *indirect testability,* because it is hard to deny to metaphysical systems a part of the verisimilitude attained by the scientific theories they help to develop.

I believe it is legitimate to go even further, such that if philosophical programs give rise to metaphysical programmes of scientific research, the scientific theories themselves can give rise, in turn, to "scientific programmes of metaphysical research". This affirmation is something more than just a speculative extrapolation. Popper's work offers convincing indications that his interest in metaphysics became, over the years, less pragmatic and more substantive, until it became one of the principal engines of his thought. Upon taking stock of the path he travelled he recognizes that problems like that of the infinite or of essentialism were at the heart of his youthful interests (Popper 1992, pp. 15–18).

Popper always positioned himself at the antipodes of the conventionalism of Duhem and Poincaré, and therefore the distinction between physics and metaphysics becomes more problematic to the degree that we approach the limits of both disciplines. The relationship between good metaphysics and good science is reciprocal, and this explains his struggles with Boltzmann and Schrödinger in regards to idealism (Popper 1992, pp. 135–138 and 156–162), with Einstein in regards to determinism (Popper 1992, pp. 127–132), and with Bohr and Heisenberg in regards to subjectivism (Popper 1982a, pp. 119–142). I will illustrate these procedures with some examples of his strident opposition to the Copenhagen Interpretation of quantum mechanics:

"When he attempted to establish atomic theory on a new basis, Heisenberg started with an epistemological programme: to rid the theory of 'unobservables', that is, of magnitudes inaccessible to experimental observation; to rid it, one might say, of metaphysical elements. [...] And this shows that Heisenberg has failed to carry through his programme. For this state of affairs only allows of two interpretations. The first would be that the particle has an exact position and an exact momentum (and therefore also an exact path) but that it is impossible for us to measure them both simultaneously. If this is so then nature is still bent on hiding certain physical magnitudes from our eyes; not indeed the position, nor yet the momentum, of the particle, but the combination of these two magnitudes, the 'position-cum-momentum', or the 'path'. This interpretation regards the uncertainty principle as a limitation of our knowledge; thus it is *subjective*. The other possible interpretation, which is an objective one, asserts that it is inadmissible or incorrect or metaphysical to attribute to the particle anything like a sharply defined '*position-cum-momentum*' or 'path': it simply has no 'path', but only either an exact position combined with an inexact momentum, or an exact momentum combined with an inexact position. But if we accept this interpretation then, again, the formalism of the theory contains metaphysical elements; for a 'path' or '*position-cum-momentum*' of the particle, as we have seen, is exactly calculable-for those periods of time during which it is in principle impossible to test it by observation. [...] Heisenberg has not so far accomplished his self-imposed task: he has not yet purged quantum theory of its metaphysical elements" (Popper 1980, pp. 217–221).

Personally, I disagree with Popper's criticism of Heisenberg, but what I would like to emphasize – and here I do agree with Popper – is that any scientist with the ambition to develop a theory has to choose between the company of a metaphysics that is good, bad or in between. However, for him or her to discard all metaphysics is just as impossible as taking away the last railroad car from a train.

It is unnecessary to search the texts of the last Popper in order to find an endorsement of the interpretation sketched out. In *The Logic of Scientific Discovery*, falsifiability is the touchstone for endorsing a theory as being "scientific". But this property is merely *de facto*: it depends on the technical capacities available in a given historical moment, or on the ability of theoreticians to find a path from its concepts to what is observable. It also depends on the degree of generality of the theory and on how far it seeks to delve into the real: the more all-embracing and profound a theory is, the more difficult it will be to put it to an empirical test. Many of these will never be able to be subjected to a crucial experiment, and will forever retain a clearly speculative character. Others will cease to be speculative beginning at a certain moment. For example, any affirmation about the dark side of the moon was "metaphysical" up until 1959 when the satellite *Luna 3* photographed it for the first time. It appears that in order to subject string theory to empirical testing, one would have to build an accelerator that would not fit within the solar system itself (Smolin 2006). I fear that this means that string theory will also be a kind of "metaphysics". With even more reason the hypothesis of multiverses will remain metaphysical as well, since nobody has been able to come up with a process of observation that would provide a valid falsification (Soler 2012).

Why, then, are articles about these notions published in journals of physics and cosmology, while articles about the agent intellect in Aristotle or Schopenhauer's Will are not? I am afraid that, apart from the possibility of applying mathematical formalisms and employing notions used in theories that have in fact been recognized as scientific, it is rather merely social-historical circumstances that have led people to situate all these theories in distinct fields of knowledge.

"Again and again suggestions are put forward-conjectures, or theories-of all possible levels of universality. Those theories which are on too high a level of universality, as it were (that is, too far removed from the level reached by the testable science of the day) give rise, perhaps, to a 'metaphysical system'. In this case, even if from this system statements should be deducible (or only semi-deducible, as foe example in the case of Spinoza's system), which belong to the prevailing scientific system, there will be no new testable statement among them; which means that no crucial experiment can be designed to test the system in question. If, on the other hand, a crucial experiment can be designed for it, then the system will contain, as a first approximation, some well corroborated theory, and at the same time also something new and something that can be tested. Thus the system will not, of course, be 'metaphysical'. In this case, the system in question may be looked upon as a new advance in the quasi-inductive evolution of science. This explains why a link with the science of the day is as a rule established only by those theories which are proposed in an attempt to meet the current problem situation; that is, the current difficulties, contradictions, and falsifications. In proposing a solution to these difficulties, these theories may point the way to a crucial experiment", (Popper 1980, p. 277).

Thomas S. Kuhn and many others have demonstrated the impossibility of using Popperian falsification in any of its versions as a univocal criterion of demarcation (Kuhn 1970, pp. 1–23). The inventors of theories tenaciously resist accepting, as good sports, the falsification of their creations, and many times reason lends them aid. This does not affect the fact that the criterion continues to be valuable, at least as a "regulative principle". Its application is only ruined if one insists on maintaining the thesis of the drastic and objective separation of science from any other enterprise of knowledge. As a result, in view of the meagre results obtained by more than two centuries of study of the problem of demarcation, the thesis in question has become an unsustainable dogma. In contrast, the progressive tendency of Popper to relativize this separation, without denying it completely, shows that his position is a more sensible one. It is odd that he was frank about recurring to shamelessly metaphysical theories precisely in order to compensate for the insufficiencies of his epistemology: "Because metaphysical realism – the idea that there is a real world to be discovered – resolves some of the problems that have remained open with my solution to the problem of induction" (Popper 1992, p. 151).

This symbiosis of the ontological and the epistemological is today the most promising alternative for escaping from the multiple dead-ends that philosophy of science has ended up in when it has sought to attain absolutely rigorous results (Feyerabend 1970, pp. 172–183). The ontological commitments assumed in books like *The Self and its Brain* and *The Open Universe* are of course debatable, but that is precisely their principal virtue. Running the risk of error is the price that has to be paid by all who wish to have a minimal opportunity of approaching the truth. Because he knew how to do so, Popper is one of the few philosophers of the 20th century who have been able to speak intimately with the great scientists of his time, about issues that went beyond mere procedural questions:

"Although I dislike the subjectivist strain in the orthodox interpretation, I am in sympathy with its rejection of the determinism of Einstein, Schrödinger and

Bohm, and with its rejection of *prima facie* deterministic theories in physics; and I agree with the substance (though hardly with the form, or with the prophetic style – the style of historical determinism) of a passage from Pauli's, taken from a letter to Born, in which Pauli rejects the deterministic research programme in the following words: "Against all retrograde efforts (Schrödinger, Bohm *et al.*, and, in a certain sense, also Einstein) I am certain that the statistical character of the *Psi*-function, and thus of the laws of nature – which you have, right from the beginning, strongly stressed in opposition to Schrödinger – will determine the style of the laws for at least some centuries. It is possible that later… something entirely new may be found, but to dream of a way back, back to the classical style of Newton-Maxwell (and it is nothing but dreams which those gentlemen indulge in), that seems to me hopeless, off the way, bad taste. And we could add 'it is not even a lovely dream'" (Popper 1982a, p. 175).

4 Final Remarks: The Uncertainty of the Frontier

I will end these reflections by noting that the frontier between scientific theories and metaphysical research programs is uncertain, which, rather than being a defect of Popper's epistemology is a wise move, because in reality the frontier that separates science and philosophy is also uncertain.

References

Arana, J. 2012. *Los sótanos del Universo. La determinación natural y sus mecanismos ocultos.* Madrid: Biblioteca Nueva.

Feyerabend, P.K. 1970. "Philosophy of Science: A Subject with a Great Past," in Stuewer, Roger H. ed. *Historical and Philosophical Perspectives of Science. Minnesota Studies in the Philosophy of Science. Vol.* v, (pp. 172–183). Minneapolis: University of Minnesota Press.

Kuhn, Th. S., 1970. "Logic of Discovery or Psychology of Research?," in Lakatos, I. and Musgrave, A. eds. *Criticism and the Growth of Knowledge,* (pp. 1–23). Cambridge Cambridge University Press.

Lakatos, I. 1978. *The Methodology of Scientific Research Programmes. Philosophical Papers. Volume 1.* Cambridge: Cambridge University Press.

Penrose, R. 1996. "La conciencia incluye ingredientes no computables," in Brockman J. ed. *La tercera cultura: más allá de la revolución científica,* (pp. 224–241). Barcelona Tusquets.

Popper, K.R. 1963. *Conjectures and Refutations. The Growth of Scientific Knowledge.* London: Routledge and Kegan Paul.

Popper, K.R. 1971. *The Open Society and Its Enemies,* 2 vols. Princeton: Princeton University Press.

Popper, K.R. 1979. *Objective Knowledge. An Evolutionary Approach.* Oxford: Clarendon Press.

Popper, K.R. 1980. *The Logic of Scientific Discovery.* London: Hutchinson.

Popper, K.R. 1982a. *Postscript to The Logic of Scientific Discovery, Vol. III. Quantum Theory and the Schism in Physics,* ed. by W.W. Bartley III. London: Hutchinson.

Popper, K.R. 1982b. *Postscript to The Logic of Scientific Discovery, Vol. II. The Open Universe. An Argument for Indeterminism,* ed. by W.W. Bartley III. London: Hutchinson.

Popper, K.R. 1983. *Postscript to The Logic of Scientific Discovery, Vol. I. Realism and the Aim of Science,* ed. by W.W. Bartley III. London: Hutchinson.

Popper, K.R. 1992. *Unended Quest. An Intellectual Autobiography.* London: Routledge.

Popper, K.R. and Kreuzer, F. 1982. *Offene Gesellschaft-offenes Universum. Ein Gespräch über das Lebenswert des Philosophen.* Vienna: F. Deuticke.

Smolin, L. 2006. *The Trouble with Physics: The Rise of String Theory, the Fall of a Science, and What Comes Next.* Boston: Houghton Mifflin Hartcourt.

Soler Gil, F.J. 2012. *Discovery or Construction? Astroparticle Physics and the Search for Physical Reality.* Frankfurt: Peter Lang.

Cognitive Problems and Practical Limits: Computers and Our Limitations

Nicholas Rescher

Abstract

Is there anything in the domain of cognitive problem solving that computers cannot manage to do? In this regard, purely theoretical limits do not represent genuine limitations in problem solving, insofar as it is no limitation to be unable to do something that cannot possibly be done. Thus, practical limits should be considered. (i) The use of inadequate information is a crucial factor here. (ii) There are practical limitations related to real-time processing difficulties as well as limitations of representation in matters of detail management. (iii) There are performative limits of prediction. They are self-insight obstacles for computers, insofar as due to its own predictive performance a computer can never function perfectly.

Consequently, there are some crucial limitations where computer determinations of computer capacity are concerned. A contrast can be made with the algorithmic decision theory, which is somewhat different. But we do here also encounter limitations by way of computer-insoluble problems. Thereafter, if we look at the human element, then humans are situated advantageously, because they can solve problems with computers. However, there are potential difficulties that this approach can overcome.

Keywords

Cognitive – practical – limits – computers – limitations – human element

1 Could Computers Overcome Our Limitations?

In view of the difficulties and limitations that beset our human efforts at answering our questions in a complex world, it becomes tempting to contemplate the possibility that computers might enable us to eliminate our cognitive disabilities and to overcome those epistemic frailties of ours. And so we may wonder: Are computers cognitively omnipotent? If a problem is to qualify as soluble at all, will computers always be able to solve it for us?

Of course, computers cannot bear human offspring, enter into contractual agreements, or exhibit heroism. But such processes address *practical* problem

 | DOI 10.1163/9789004325401_004

relating to the management of the affairs of human life and so do not count in the present cognitive context. Then too we must put aside *evaluative* problems of normative bearing or of matters of human affectivity and sensibility: computers cannot offer us meaningful consolation or give advice to the lovelorn. The issue presently at hand regards the capacity of computers to resolve *cognitive* problems of the regarding matters of empirical or formal fact. Typically, the sort of problems that will concern us here are those that characterize the sciences, in particular problems relating to the description, explanation, and prediction of the things, events, and processes that comprise the realm of physical reality. And to all visible appearances computers are ideal instruments for handling the matters of cognitive complexity that arise in such contexts. The question, then, is: Is there anything in the domain of cognitive problem solving that computers cannot manage to do?

The history of computation in recent times is one of a confident march from triumph to triumph. Time and again, those who have affirmed the limitedness of computers have been forced into ignominious retreat as increasingly powerful machines implementing increasingly ingenious programs have been able to achieve the supposedly unachievable. However, the question on the present agenda is not "Can computers *help* with problem-solving?" – an issue that demands a resounding affirmative and needs little further discussion. There is no doubt whatever that computers can do a lot here – and very possibly more than we ourselves can. But there is an awesomely wide gap between a lot and *everything*.

First some important preliminaries. To begin with, we must, in this present context, recognize that much more is at issue with a "computer" than a mere electronic calculating machine understood in terms of its operational hardware. For one thing, software also counts. And, for another, so does data acquisition. As we here construe computers, they are electronic information-managing devices equipped with data banks and augmented with sensors as autonomous data access. Such "computers" are able not only to *process* information but also to *obtain* it. Moreover, the computers at issue here are, so we shall suppose, capable of discovering and learning, and thereby able significantly to extend and elaborate their own initially programmed modus operandi. Computers in this presently operative sense are not mere calculating machines, but general problem solvers along the lines of the fanciful contraptions envisioned by the aficionados of artificial intelligence. These enhanced computers are accordingly question-answering devices of a very ambitious order.

On this expanded view of the matter, we must also correspondingly enlarge our vision both of what computers can do and of what can reasonably be asked of them. For it is the potential of computers as an instrumentality for universal problem solving (UPS) that concerns us here, and not merely their more limited role in the calculations of algorithmic decision theory (ADC). The computers at issue will thus be prepared to deal with factually substantive as well as merely formal (logico-mathematical) issues. And this means that the questions we can ask are correspondingly diverse. For here, as elsewhere, added power brings added responsibility. The questions it is appropriate to ask thus can relate not just to matters of calculation but to the things and processes of the world.

Moreover, some preliminary discussion of the nature of "problem solving" is required because one has to become clear from the outset about what it is to *solve* a cognitive problem. Obviously enough, this is a matter of answering questions. Now "to answer" a question can be construed in three ways: to offer a *possible* answer, to offer a *correct* answer, and finally to offer a *credible* answer. It is the third of these senses that will be at the center of concern here. And with good reason. For consider a problem solver that proceeds in one of the following ways: it replies "yes" to every yes/no question; or it figures out the range of possible answers and then randomizes to select one; or it proceeds by "pure guesswork".

Even though these so-called "problem solvers" may give the correct response some or much of the time, they are systematically unable to resolve our questions in the presently operative credibility-oriented sense of the term. For the obviously sensible stance calls for holding that *a cognitive problem is resolved only when an appropriate answer is convincingly provided* – that is to say, when we have a solution that we can responsibly accept and acknowledge as such. Resolving a problem is not just a matter of having an answer, and not even of having an answer that happens to be correct. The actual resolution of a problem must be credible and convincing – with the answer provided in such a way that its cogency is recognizable. In general problem solving we want not just a dictum but an *answer* – a response equipped with a contextual rationale to establish its credibility in a way accessible to duly competent recipients. To be warranted in accepting a third-party answer we must ourselves have case-specific reasons to acknowledge it as correct. A response whose appropriateness as such cannot secure rational confidence is no answer at all.[1]

With this crucial preliminary out of the way, we are ready to begin.

---

1 The salient point is that unless I can *tell* (i.e., myself be able to claim justifiedly) that you a justified in your claim, I have no adequate grounds to accept it: it has not been made credib to me, irrespective of how justified you may be in regard to it. To be sure, for your claim to b credible for me I need not know *what* your justification for it is, but I must be in a positic to realize *that* you are justified. Your reasons may even be incomprehensible to me, but f

General-Principle Limits are Not Meaningful Limitations

'he question before us is: "Are there *any* significant cognitive problems that omputers cannot solve?" Now it must be acknowledged from the outset that ertain problems are inherently unsolvable in the logical nature of things. One annot square the circle. One cannot co-measure the incommensurable. One annot decide the demonstrably undecidable nor prove the demonstrably unrovable. Such tasks represent absolute limitations whose accomplishment is heoretically impossible – unachievable for reasons of general principle rooted n the nature of the realities at issue.[2] And it is clear that inherently unsolvable roblems cannot be solved by computers either.[3]

Other sorts of problems will not be unsolvable as such but will, neverthe-ess, be demonstrably prove to be computationally intractable. For with re-pect to *purely theoretical* problems it is clear from Turingesque results in algo-ithmic decision theory (ADT) that there will indeed be computer insolubilia - mathematical questions to which an algorithmic respondent will give the vrong answer or be unable to give any answers at all, no matter how much ime is allowed.[4] But this is a mathematical fact which obtains of necessity o that this whole issue can be also set aside for present purposes. For in the resent context of universal problem solving (UPS) the necessitarian facts of Gödel-Church-Turing incompleteness become irrelevant. Here any search for *neaningful* problem-solving limitations will have to confine its attention to roblems that are in principle solvable: *demonstrably* unsolvable problems are eside the point of present concern because an inability to do what is in prin-iple impossible hardly qualifies as a limitation, seeing that it makes no sense o ask for the demonstrably impossible.

For present purposes, then, it is limits of *capability* not limits of *feasibility* hat matter. In asking about the problem-solving limits of computers we are ooking to problems that *computers* cannot resolve but that other problem

---

credibility I require a rationally warranted assurance – perhaps only on the basis of general principles – that those reasons are both extant and cogent.

2 On unsolvable calculating problems, mathematical completeness, and computability see Davis (1982). See also Pour-El and Richards (1989) or on a more popular level Hofstadter (1979).

3 Some problems are not inherently unsolvable but cannot in principle be settled by computers. An instance is "What is an example of word that no computer will ever use?" Such problems are inherently computer-inappropriate and for this reason a failure to handle them satisfactorily also cannot be seen as a meaningful limitation of computers.

4 On Gödel's theorem see Shanker (1988), a collection of essays that provide instructive, personal, philosophical, and mathematical perspectives on Gödel's work.

solvers conceivably can. The limits that will concern us here are accordingly not rooted in conceptual or logico-mathematical infeasibilities of general principle nor in absolute physical impossibilities, but rather in performatory limitations imposed specifically upon computers by the world's contingent *modus operandi*.

And in this formulation the adverb "specifically" does real work by way of ruling out certain computer limitations as irrelevant. Things standing as they do, some problems will simply be too large given the inevitable limitations on computers in terms of memory, size, processing time, and output capacity. Suppose for the moment that we inhabit a universe which, while indeed boundless, is nevertheless finite. Then no computer could possibly solve a problem whose output requires printing more letters or numbers than there are atoms in the universe. Such problems ask computers to achieve a task that is not "substantively meaningful" in the sense that no physical agent at all – computer, organism, or whatever, could possibly achieve it. The problems that concern us here are those that are not solution-precluding on the basis of inherent mathematical or physical impossibilities. To reemphasize: our concern is with the performative limitations of computers with regard to problems that are not inherently intractable in the logical or physical nature of things.

3 Practical Limits: Inadequate Information

Often the information needed for credible problem-resolution is simply unavailable. Thus no problem-solver can at this point in time provide credible answers to questions like "What did Julius Caesar have for breakfast on that fatal Ides of March?" or "Who will be the first auto accident victim of the next millennium?" The information needed to answer such questions is just not available at this stage. In all problem-solving situations, the performance of computers is decisively limited by the quality of the information at their disposal. "Garbage in, garbage out," as the saying has it. But matters are in fact worse than this. Garbage can come out even where no garbage goes in.

One clear example of the practical limits of computer problem-solving arises in the context of prediction. Consider the two prediction problems set out in Display 1. On first sight, there seems to be little difficulty in arriving at a prediction in these cases. But now suppose that we acquire some further data to enlarge our background information: pieces of information supplementary to – but nowise conflicting with or corrective of – the given premisses:

Case 1: *X* is extremely, indeed *inordinately* fond of Trollope.
Case 2: *Z* also promised to repay his other neighbor the $7.00 he borrowed on the same occasion.

Display 1

Case 1
Data: *X* is confronted with the choice of reading a novel by Dickens or one by Trollope. And further: *X* is fond of Dickens.
Problem: To predict which novel *X* will read.
Case 2
Data: *Z* has just exactly $10.00. And further: *Z* promised to repay his neighbor $7.00 today. Moreover, *Z* is a thoroughly honest individual.
Problem: To predict what *Z* will do with his money.

Note that in each case our initial information is nowise abrogated but merely enlarged by the additions in question. But nevertheless in each case we are impelled, in the light of that supplementation, to *change* the response we were initially prepared and rationally well advise to make. Thus when I know nothing further of next year's Fourth of July parade in Centerville U.S.A., I shall predict that its music will be provided by a marching band; but if I am additionally informed that the Loyal Sons of Old Hibernia have been asked to provide the music, then bagpipes will now come to the fore.

It must, accordingly, be recognized that the search for rationally appropriate answers to certain questions can be led astray not just by the *incorrectness* of information but by its *incompleteness* as well. The specific body of information that is actually at hand is not just important for problem resolution, it is *crucial*. And we can never be unalloyedly confident of problem-resolutions based on incomplete information, seeing that further information can always come along to upset the applecart. As available information expands, established problem-resolutions can always become destabilized. One crucial practical limitation of computers in matters of problem solving is thus constituted by the inevitable incompleteness (to say nothing of potential incorrectness) of the information at their disposal. And here the fact that computers can only ever ingest finite – and thus incomplete – bodies of information means that their problem-resolving performance is always at risk. (Moreover, this sort of risk exists quite apart from others, such as the fact that computerized problem-resolutions are always the product of many steps, each of which involves a nonzero probability of error.) If we are on a quest for certainty, computers will not help us to get there.

4 Practical Limits: Transcomputability and Real-Time Processing Difficulties

Apart from *uncomputable* (computationally irresolvable) problems there is also the range of *transcomputable* problems, problems whose computational requirements exceed the physical bounds and limits that govern the concrete realization of theoretically designed algorithmic machines.[5] Because computers are physical devices, they are subject to the laws of physics and limited by the realities of the physical universe. In particular, since a computer can process no more than a fixed number of bits per second per gram, the potential complexity of algorithms means that there is only so much that a given computer can possibly manage to do.

Then there is also the temporal aspect. To solve problems about the real world, a computer must of course be equipped with information about it. But securing and processing information is a time-consuming process and the time at issue can never be reduced to an instantaneous zero. Time-constrained problems that are enormously complex – those whose solution calls for securing and processing a vast amount of data – can exceed the reach for any computer. At some point it always becomes impossible to squeeze the needed operations into available time. There are only so many numbers that a computer can crunch in a given day. And so if the problem is a predictive one it could find itself in the awkward position that it should have started yesterday on a problem only presented to it today. Thus even under the (fact-contravening) supposition that the computer can answer *all* of our questions, it cannot, if we are impatient enough, produce those answers as promptly as we might require them. Even when given, answers may be given too late.

5 Practical Limits: Limitations of Representation in Matters of Detail Management

This situation is emblematic of a larger issue. Any computer that we humans can possibly contrive here on earth is going to be finite: its sensors will be finite, its memory (however large) will be finite, and its processing time (however fast) will be finite.[6] Moreover, computers operate in a context of finite instructions and finite inputs. Any representational model that functions by

5 On this issue see Brennerman (1977, pp. 167–174).

6 For a comprehensive survey of the physical limitations of computers see Leiber (199[illegible] pp. 23–54).

means of computers is of finite complexity in this sense. It is always a finitely characterizable system: Its descriptive constitution is characterized in finitely many information-specifying steps and its operations are always ultimately presented by finitely many instructions. And this array of finitudes means that a computer's modelling of the real will never capture the inherent ramifications of the natural universe of which it itself is but a minute constituent. Artifice cannot replicate the complexity of the real; reality is richer in its descriptive constitution and more efficient in its transformatory processes than human artifice can ever manage to realize. For nature itself has a complexity that is effectively endless, so that no finistic model that purports to represent nature can ever replicate the detail of reality's make-up in a fully comprehensive way, even as no architect's blueprint-plus-specifications can possibly specify *every* feature of the structure that is ultimately erected. In particular, the complications of a continuous universe cannot be captured completely via the resources of discretized computer languages. All endeavors to represent reality – computer models emphatically included – involve some element of oversimplification, and in general a great deal of it.

The fact of the matter is that reality is too complex for adequate cognitive manipulation. Cognitive friction always enters into matters of information management – our cognitive processing is never totally efficient, something is always lost in the process; cognitive entropy is always upon the scene. But as far as knowledge is concerned, nature does nothing in vain and so encompasses no altogether irrelevant detail. Yet Oversimplification always makes for losses, for deficiencies in cognition. For representational omissions are never totally irrelevant, so that no oversimplified descriptive model can get the full range of predictive and explanatory matters exactly right. Put figuratively, it could be said that the only "computer" that can keep pace with reality's twists and turns over time is the universe itself. It would be unreasonable to expect any computer model less complex than this totality itself to provide a fully adequate representation of it, in particular because that computer model must of course itself be incorporated *within* the universe.

6 Performative Limits of Prediction – Self-Insight Obstacles

Another important sort of practical limitation to computer problem-solving arises not from the inherent intractability of questions but from their unsuitability for particular respondents. Specifically, one of the issues regarding which a computer can never function perfectly is its own predictive performance. One critical respect in which the self-insight of computers is limited arises in connection with what is known as "the Halting Problem" in algorithmic decision theory (ADC). Even if a problem is computer solvable – in the sense that a suitable computer will demonstrably be able to find a solution by keeping at it long enough – it will in general be impossible to foretell how long a process of calculation will actually be needed. There is not – and demonstrably cannot be – a *general* procedure for foretelling with respect to a particular computer and a particular problem: "Here is how long it will take to find the solution – and if the problem is not solved within this timespan then it is not solvable at all." No computer can provide general insight into how long it – or any other computer, for that matter – will take to solve problems. The question "How long is long enough?" demonstrably admits of no general solution here.

And computers are – of necessity! – bound to fail even in much simpler self predictive matters. Thus consider confronting a predictor with the problem posed by the question:

P1: When next you answer a question, will the answer be negative?

This is a question which – for reasons of general principle – no predictor can ever answer satisfactorily.[7] For consider the available possibilities:

Answer given	*Actually correct answer*	*Agreement?*
YES	NO	NO
NO	YES	NO
CAN'T SAY	NO	NO

On this question, there just is no way in which a predictive computer's response could possibly agree with the actual fact of the matter. Even the seemingly plausible response "I can't say" automatically constitutes a self-falsifying answer, since in giving this answer the predictor would automatically make "No" into the response called for by the proprieties of the situation.

Here, then, we have a question that will inevitably confound any conscientious predictor and drive it into baffled perplexity. But of course the problem poses a perfectly meaningful question to which *another* predictor could give a putatively correct answer – namely, by saying: "No – that predictor cannot answer this question at all; the question will condemn a predictor (Predictor No. 1) to baffled silence." But of course the answer "I am responding with baffle

7 As stated this question involves a bit of anthropomorphism in its use of "you". But this is s only for reasons of stylistic vivacity. That "you" is, of course, only shorthand for "comput number such-and-such".

silence" is one which that initial predictor cannot cogently offer. And as to that baffled silence itself, this is something which, as such, would clearly constitute a defeat for Predictor No. 1. Still, that question which impelled Predictor No. 1 into perplexity and unavoidable failure presents no problem of principle for Predictor No. 2. And this clearly shows that there is nothing improper about that question as such. For while the question posed in P_1 will be irresolvable by a *particular* computer, and so it could – in theory – be answered by *other* computers is not irresolvable by computers-in-general.

However, there are other questions that indeed are computer insolubilia for computers-at-large. One of them is:

P2: What is an example of a predictive question that no computer will ever state?

In answering *this* question the computer would have to stake a claim of the form: "*Q* is an example of a predictive question that no computer will ever state." And in the very making of this claim the computer would falsify it. It is thus automatically unable to effect a satisfactory resolution. However, the question is neither meaningless nor irresolvable. A *non-computer* problem solver could in theory answer it correctly. Its presupposition, "There is a predictive question that no computer will ever consider" is beyond doubt true. What we thus have in P_2 is an example of an in-principle solvable – and thus "meaningful" – question which, as a matter of necessity in the logical scheme of things, no problem-solving computer can ever resolve satisfactorily. The long and short of it is that every predictor – computers included – is bound to manifest versatility-incapacities with respect to its own predictive operations.[8]

However, from the angle of our present considerations, the shortcoming of problems P_1 and of P_2 is that they are computer irresolvable on the basis of theoretical general principles. And it is therefore not appropriate, on the present perspective – as explained above – to count this sort of thing as a computer limitation. Are there any other, less problematic examples?

7 Performative Limits: A Deeper Look

At this point we must contemplate some fundamental realities of the situation confronting our problem-solving resources. The first of these is that no computer can ever reliably determine that all its more powerful compeers

8 On the inherent limitation of predictions see Rescher (1998).

are unable to resolve a particular substantive problem (that is, one that is inherently tractable and not demonstrably unsolvable on logico-conceptual grounds). And this means that:

T_1: *No computer can reliably determine that a given substantive problem is altogether computer irresolvable.*

This is to say that no computer can reliably determine that a particular substantive problem *p* is such that no computer can resolve it: $(\forall C) \sim C$ res p. We thus have:

$$\sim(\exists C')(\exists P)C' \text{ det } (\forall C)\sim C \text{ res } P$$

or equivalently

$$(\forall C')(\forall P)\sim C' \text{ det } (\forall C)\sim C \text{ res } P$$ [9]

A brief explanation is needed regarding the use of "determine" that is operative here. In the present context this is a matter of so functioning as to be able to secure rational conviction for the claim at issue. As was emphasized above, we want not just answers but *credible* answers.

Moreover, something that we are not prepared to accept from any computer is cognitive megalomania. No computer is, so we may safely suppose, ever able to achieve credibility in staking a claim to the effect that no substantive problem whatever is beyond the capacity-reach of computers. And this leads to the thesis:

T_2: *No computer can reliably determine that all substantive problems whatever are computer resolvable.*

9 Note that T_1 is weaker than:

T_3: No computer can reliably determine that there are substantive problems that are computer irresolvable:

$$\sim(\exists C')\, C' \text{ det } (\exists P)(\forall C) \sim C \text{ res } P$$

This stronger thesis is surely false, but the truth of the weaker T_1 nevertheless remains intact.

The falsity of T_3 can be shown by establishing its denial: $(\exists C')\, C'$ det $(\exists P)(\forall C) \sim C$ res P. First note that in a world of computers whose capabilities are finite there is for each computer *Ci* some problem *Pi* that *Ci* cannot resolve. The conjunctive compilation of all these problems, namely *P*x will thus lie beyond the capacity of all computers. So $(\forall C) \sim C$ res Px and therefore $(\exists P)(\forall C) \sim C$ res P. Now the reasoning by which we have just determined this can itself clearly be managed by a computer, and thus $(\exists C')\, C'$ det $(\exists P)(\forall C) \sim C$ res P. Q.E.D.

That is to say that no computer can convincingly establish that whenever a substantive problem *p* is at issue, then some computer can resolve it – in other words, that for any and every substantive problem *p*: $(\exists C)$ *C* res *p*. Thus:

$$\sim(\exists C')\ C' \text{ det } (\forall P)(\exists C)\ C \text{ res } P$$

or equivalently:

$$(\forall C')\sim C' \text{ det} \sim(\exists P)(\forall C)\sim C \text{ res } P$$

Neither can a computer reliably determine that an arbitrarily given substantive problem is computer irresolvable (T_1) nor can it reliably determine that no such problem is computer irresolvable (T_2).

We shall not now expatiate upon the rationale of these theses. This issue of establishing their plausibility – whose pursuit would at this point unduly interrupt the flow of present deliberations – will be postponed until the Appendix. All that matters at this juncture is that the principles in question merit acceptance – and do so not as a matter of abstractly mathematico-logical considerations, but owing to the world's practical realities.

The relationship between theses T_1 and T_2 comes to light more clearly when one considers their formal structure. The claims at issue are as follows:

T_1: For all C': $(\forall P)\sim C'$ det $(\forall C)\sim C$ res P
T_2: For all C': $\sim C'$ det $(\forall P)\sim(\forall C)\sim C$ res P

Now let us also adopt the following two abbreviations

- *C*-un *p* for: $\sim C$ det *P* ("*C* is unable to determine that *p*")
- *X*(*p*) for: $(\forall C)\sim C$ res *P* ("*p* is computer-unresolvable")

Then

T_1 = For all *C*: $(\forall P)C$-un $X(P)$
T_2 = For all *C*: *C*-un $(\forall P)\sim X(P)$

As this makes clear, both theses alike indicate a universal computer incapacity in relation to computer-unresolvability theses of the form $(\forall C)\sim C$ res *p*. Thus T_1 and T_2 both reflect ways in which computers encounter difficulty in obtaining a credible grip on such universal incapacity. Fixing the bounds of computer solvability is beyond the capacity of any computer.

It should be noted that in his later writings, Kurt Gödel himself took a lir regarding mathematics analogous to that which the present discussion tak with respect to general problem solving. For he maintained that no single pa ticular axiomatic proof-systematization will be able to achieve universali with respect to provability in general.[10] And so, even as Gödel saw algorithn as inherently incapable of doing full justice to mathematics, so the prese argumentation has it that problem-solving computers cannot do full justice science. Both theses alike implement the common idea that, notwithstandir the attractions and advantages of rigorous reasoning, the fact remains that in complex world it is bound to transpire that truth is larger than rigor.

8 Contrast with Algorithmic Decision Theory

The unavailability of a universal problem solver in the setting of gener problem solving has far-reaching theoretical implications. For it means th in universal problem solving (UPS) we face a situation regarding the capab ity of computers that is radically different from that of algorithmic decisio theory (ADC).

In algorithmic decision theory we have Church's Thesis:

> Wherever it is possible for computation to decide an issue, this resolutio can be achieved by means of effective calculation. Thus computation resolvability/decidability (an informal conception!) can to all useful ir tents and purposes be equated with algorithmically effective compu ability (which is rigorously specifiable):[11]
>
> (C) $\text{sol}\ P \rightarrow (\exists C)(C\ \text{res}\ P)$

To this thesis one can adjoin Alan Turing's great insight that there can be "universal computer" (a "Turing machine")[12] – a device that can solve a cor putational problem if any calculating machine can:

(T) $(\exists C)(C\ \text{res}\ P) \rightarrow \text{T res}\ P$

---

10 See Feferman (1995), and especially the 1951 paper on "Some basic theorems on the fou dation of mathematics and their indications."

11 On Church's Thesis see Rogers (1967) and Davis (1965).

12 On Turing machines see Hesken (1988).

Combining these two theses, we arrive at the result that in the sphere of algorithmic computation solvability-at-large is tantamount to resolvability by a Turing machine:

(M) sol $P \rightarrow$ T res P

Here one machine can speak for the rest: If a problem is resolvable at all by algorithmic calculations, then a Turing machine can resolve it. In algorithmic decision theory (ADC), there is thus an absolute, across-the-board conception of solvability.

But when we turn our perspective to universal problem solving (UPS) this monolithic situation is lost. Here the state of things is no longer Turingesque: there is not and cannot be a universal problem solver.[13] As we have seen, for any problem-solver there will automatically be some correlatively unsolvable problems – problems that it cannot resolve but others can – along the lines of the aforementioned computer-embarrassing question P_1 ("Will the next answer that you give be negative?"). Once we leave the calm waters of algorithmic computation and venture into the turbulent sea of problem-solving computation in general, it becomes impracticable for any computer to survey all the possibilities. Here the overall range of computer-resolvable problems extends beyond the information horizon (the "range of vision" so to speak) of any given computer, so that no computer can make convincing claims about this range as a whole. In particular, these deliberations mean we would not – and should not – be prepared to take a computer's word for it if it stakes a claim of the format "*Q* is a (substantive) question that no computer whatsoever could possibly resolve."

9 A Computer Insolubilium

The time has come to turn from generalities to specifics. At this point we can confront a problem-solving computer (*any* such computer) with the challenging question:

13 But could one not simply connect computers up with one another so as to create one vast mega-computer that could thereby do anything that any computer can. Clearly this can (in theory) be done when the set of computers at issues includes "all currently existent computers". But of course we cannot throw future ones into the deal, let alone merely possible ones.

P3: What is an example of a (substantive) problem that no computer whatsoever can resolve?

There are three possibilities here:

1. The computer offers an answer of the format "*P* is an example of a problem that no computer whatsoever can resolve." For reasons already canvassed we would not see this as an acceptable resolution, since by T_1 our respondent cannot achieve credibility here.

2. The computer responds: "No can do: I am unable to resolve this problem it lies outside my capability." We could – and would – accept this response and take our computer at its word. But the response of course represents no more than computer acquiescence in computer incapability.

3. The computer responds: "I reject the question as being based on an inappropriate presupposition, namely that there indeed are problems that no computer whatsoever can resolve." We ourselves would have to reject this position as inappropriate in the face of T_2. The response at issue here is one that we would simply be unable to accept at face value from a computer.

It follows from such deliberations that P_3 is itself a problem that no computer can resolve satisfactorily.

At this point, then, we have realized the principal object of the discussion. We have been able to identify a meaningful concrete problem that is computer irresolvable for reasons that are embedded – via theses T_1 and T_2 – in the world's empirical realities. For – to reemphasize – our present concern is with issues of general problem solving and not algorithmic decision theory.

10 The Human Element: Can People Solve Problems that Computers Cannot?

Our discussion has not, as yet, entered the doctrinal terrain of discussions along the lines of Hubert L. Dreyfus' *What Computers Still Can't Do* (1992).[14] For the project that is at issue there is to critique the prospects of "artificial intelligence" by identifying processes involving human intelligence and behavior that computers cannot manage satisfactorily. And they accordingly compare computer information processing with human performance in an endeavor to show that there are things that humans can do that computers cannot

14 This book is an updated edition of his earlier *What Computers Can't Do* (1972).

accomplish. However, the present discussion has to this point proceeded with a view solely to problems that computers cannot manage to resolve. Whether *humans* can or cannot resolve them is an issue that has remained out of sight.

And so there is a big question that yet remains untouched, namely: Is there any sector of this problem-solving domain where the human mind enjoys a competitive advantage over computers? Or does it transpire that wherever computers are limited, humans are always limited in similar ways?

In addressing this issue, let us be precise about the question that now faces us. It is:

P4: Are there problems that computers cannot solve satisfactorily but people can?

And in fact what we would ideally like to have is not just an abstract answer to P_4, but a concrete answer to:

P5: What is an example of a problem that computers cannot solve satisfactorily but people can?

What we are now seeking is a computer-defeating question that has the three characteristics of (i) posing a meaningful problem, (ii) being computer-unsolvable, and (iii) admitting of a viable resolution by intelligent non-computers, specifically humans.[15]

This, then is what we are looking for. And – lo and behold! – *we have already found it.* All we need do is to turn around and look back to P_3. After all, P_3 is – so it was argued – a problem that computers cannot resolve satisfactorily, and this consideration automatically provides us – people that we are – with the example that is being asked for. In presenting P_3 within its present context we have in fact resolved it. And moreover P_5 is itself also a problem of just this same sort. It too is a computer-unresolvable question that people can manage to resolve.[16]

---

15 For some discussion of this issue from a very different point of approach see Penrose (1989).

16 We have just claimed P_5 as computer irresolvable. And this contention, of course, entails $(\exists P)(\forall C){\sim}C$ res P or equivalently ${\sim}(\forall P)(\exists C)C$ res P. Letting this theses be T_3, we may recall that T_2 comes to $(\forall C){\sim}C$ det ${\sim}T_3$. If T_3 is indeed true, then this contention – that is T_2 – will of course immediately follow.

In the end, then, the ironic fact remains that the very question we are considering regarding cognitive problems that computers cannot solve but people can provides its own answer.[17] P_3 and P_5 appear to be eligible for membership in the category of "academic questions" – questions that are effectively self-resolving – a category which also includes such more prosaic members as: "What is an example of a question formulated in English?" and "What is an example of a question that asks for an example of something?" The presently operative mode of computer unsolvability thus pivots on the factor of self-reference – just as is the case with Gödelian incompleteness.

To be sure, their inability to answer the question "What is a question that no computer can possibly resolve?" is – viewed in a suitably formidable perspective – a token of the power of computers rather than of their limitedness. After all, we see the person who maintains "I can't think of something I can't accomplish" not as unimaginative but as a megalomaniac – and one who uses "we" instead of "I" as only slightly less so. But nevertheless, in the present case this pretension to strength marks a point of weakness.

The key issue is whether computers might be defeated by questions which other problem solvers, such as humans, could overcome. The preceding deliberations indicate that there indeed are such limitations. For the ramifications of self-reference are such that one computer could satisfactorily answer certain questions regarding the limitation of the general capacity of computers to solve questions. But humans can in fact resolve such questions because, with them, no self-reference is involved.

11 Potential Difficulties

The time has now come for facing up to some possible objections and difficulties.

An objection that deserves to be given short shrift runs as follows: "But the sort of computer insolubilium represented by self-reference issues like those

---

17 Someone might suggest: "But one can use the same line of thought to show that there are computer-solvable problems that people cannot possibly solve by simply interchanging the reference to 'computers' and 'people' throughout its preceding argumentation?" But this will not do. For the fact that it is people that use computers means that one can credit people with computer-provided problem solutions via the idea that *people can solve problems with computers*. But the reverse cannot be claimed with any semblance of plausibility. The situation is not in fact symmetrical and so the proposed interchange will not work. This issue will be elaborated in the next section.

of P_5 is really not the kind of thing I was expecting when contemplating the title of the paper." Expectations do not really count for much in this sort of situation. After all, nobody faults Gödel for not having initially coming up with sort of examples that people might have expected regarding the incompleteness of formalized arithmetic – some insoluble diophantine problem in number theory.[18]

But of course other objections remain.

For example, do those instanced problems really lie outside the domain of trustworthy computer operation? Could computers not simply follow in the wake of our own reasoning here and instance P_3 and P_5 as self-resolving? Not really. For in view of the considerations adduced in relation to T_1-T_2 above, a computer cannot convincingly monitor the range of computer-tractable problems. And so, the responses of a computer in this sort of issue simply could not secure rational conviction.

But what if a computer were to establish its reliability regarding such supposedly "computer irresolvable" questions indirectly? What about a reliable black box? Could a computer not acquire reliability simply by compiling a good track record?

Well...yes and no. A black box can indeed establish credibility by compiling a good track record of correct responses. But it can do so only when this track record is issue-homogeneous with the matter at hand: when those correct responses relate to questions of the same type as the one that is at issue. Credibility is not transferable from botany to mathematics or from physics to theology. The only productively meaningful track record would have to be one complied *in a reference class of similar cases.* Now just how does type-homogeneity function with respect to our problem? What is the "type of problem" that is at issue here? The answer is straightforward: it is *questions that for reasons of principle qualify as being computer-intractable.* But how could a computer establish a good track record here? Only by systematically providing the responses we can reasonably deem to be correct on wholly independent grounds.

The situation that arises here would thus be analogous to that of a black box that systematically forecasts tomorrow's headlines correctly. This sort of thing is indeed imaginable – it is a logically feasible possibility (and thereby, no doubt, an actuality in the realm of science fiction). But we would be so circumstanced as to deem that black box's performance as miraculous. And we

18 Such examples were to come along only later with the relating of Gödel's results to number-theoretic issues relating to the solution of Diophantine equations. For a good expository account see Davis and Hersh (1978, pp. 554–571). (I owe this reference to Kenneth Manders.)

do not – cannot – accept this as a *practical* possibility for the real world. It is a fanciful hypothesis that we would reject out of hand until such time as actual circumstances confronted us with its realization – a prospect we would dismiss as utterly unrealistic. It represents a bridge that we would not even think about crossing until we actually got there – simply because we have a virtually ineradicable conviction that actually getting there is something that just will not happen.

"But surely the details of this discussion are not so complex that a computer capable of defeating grandmasters at chess could not handle them as well." There thus still remains a subtle and deep difficulty. One recent author formulated the problem as follows: "In a way, those who argue for the existence of tasks performable by people and not performable by computers are forced into a position of never-ending retreat. If they can specify just what their task involves, then they [must] admit the possibility of programming it on some machine. … [And] even if they construct proofs that a certain class of machines cannot perform certain tasks, they are vulnerable to the possibility of essentially new classes of machines being described or built."[19]

After all, when people can solve a certain problem successfully, then they can surely "teach" this solution to a computer by simply adding the solution to its information-store. And thereafter the computer can also solve the problem by replicating the solution – or if need be by simply looking it up.

Well, so be it. It is certainly possible for a computer to maintain a solution registry. Every time some human solves a problem somewhere somehow, it is duly entered into this register. And then the computer can determine the person-solvability of a given problem by the simple device of going and "looking it up". But this tactic represents a hollow victory. First of all, it would give the computer access only to person-resolved problems and not to person-resolvable ones. But, more seriously yet, if a computer needs *this* sort of input for answering a question, then we could hardly characterize the problem at issue as computer solvable in any but a Pickwickian sense.

At this point the issue of the scoring system becomes crucial. We now confront what is perhaps the most delicate and critical part of the inquiry. For now we have to face and resolve the conceptual question of how the attribution of credit works in matters of problem solving.

Clearly if *all* of the inferential steps essential to establishing a problem solution as such were computer-performed, and *all* of the essential data inputs were computer-provided, then computers will have to be credited with that problem solution. But what of the mixed cases where some essential contributions were made on both sides – some by computers and some by

19 Weinberg (1967, p. 173).

people. Here the answer is that *credit for mixed solutions lies automatically with people*. For if a computer "solves" the problem in a way that is overtly and essentially dependent on people-provided aid, then its putative "solution" can no longer count as authentically computer-provided. A problem that is "not solvable by persons alone but yet solvable when persons are aided by computers" is still to be classed as person solvable, while a problem that is "not solvable by computers alone but yet solvable when computers are aided by persons" is not to be classed as computer solvable. (Or at any rate is not so until we reach the science-fiction level of self-produced, self-programmed, independently evolving computers that manage to reverse the master-servant relationship here.) For as matters stand, the scoring system used in these matters is simply not "fair". The seemingly table-turning question "Is there a problem that people cannot solve but computers can?" automatically requires a negative response once one comes to realize that *people can and do solve problems with computers*.[20] The conception of "computer-provided solutions" works in such a way that here computers must not only do the job but actually *the whole job*. And on this basis the difficulty posed by that subtle objection can be dismissed.

The crucial point is that while people use computers for problem solving, the converse simply does not hold: the prospect that computers solve problems by using people as investigative instrumentations is unrealistic – barring a technologico-cultural revolution that sees the emergence of functionally autonomous computers, operative on their own and able to press people into their service.

Does such a principle of credit allocation automatically render people superior to computers? Not necessarily. Quite possibly the things that computers cannot accomplish in the way of problem solving are things people cannot accomplish either – be it with or without their means. The salient point is surely that much of what we would ideally like to do, computers cannot do either. They can indeed diminish but cannot eliminate our limitations in solving the cognitive problems that arise in dealing with a complex world, that is, in effect, a realm where *every* problem-solving resource faces some ultimately insuperable obstacles.[21]

20 To be sure, there still remains the question: "Are there problems that people can solve *only* with the aid of computers?" But the emphatically affirmative answer that is virtually inevitable here involves no significant insult to human intelligence. After all, the same concession must be made with regard to reference aids of all sorts, the telephone directory being an example.

21 I am grateful to Gerald Massey, Laura Ruetsche, and Alexander Pruss for constructive comments and useful suggestions on a draft of this paper.

Appendix

On the Plausibility of T_1 and T_2

The task of this Appendix is to set out the plausibility considerations that establish the case for accepting the pivotal theses T_1 and T_2.

A helpful starting point for these deliberations is provided by the recognition that the inherently progressive nature of pure and applied science ensures the prospect of continual improvements in the development of ever more capable instrumentalities for general problem solving. No matter how well we are able to do at any given state-of-the-art stage in this domain the prospect of further improvements always lies open. Further capabilities in point of information access and/or processing capacity can always be added to any realizable computer, no matter how powerful it may be. And this suffices to substantiate the realization that there is no inherent limit here: for every particular problem solver that is actually realized there is (potentially) some other whose performative capability is greater.[22]

Now it lies in the nature of things that in cognitive matters, an agent possessed of a lesser range of capabilities will always underperform one possessed of a greater range. More powerful problem solvers can solve more problems. A chess player who can look four moves ahead will virtually always win out over one who can manage only three moves. A crossword puzzlist who can manage words of four syllables will surpass one who can manage only three. A mathematician who has mastered the calculus will outperform one whose competency is limited to arithmetic. And this sort of thing holds for genera problem solving as well.

22 To be sure, due care must be taken in construing the idea of greater performative capabil ity. Thus let *C* be the computer in question and let *s* be a generally computer-undecidabl statement. Then consider the question:

(*Q*) Is it the case that at least one of the following two conditions holds? (i) You (i.e the computer at work on the question *Q* now being posed) are *C*, (ii) *s* is true.

Since (i) obtains, *C* can answer this interrogation affirmatively. But since *s* is (by hy pothesis) computer-undecidable, no other computer whatever can resolve *Q*. It migh thus appear that no computer whatever could have greater general capability than ar other. However, its essential use of "you" prevents *Q* actually qualifying as "a question tha *C* can answer but *C'* cannot." For in fact *different questions* are being posed – and differer problems thus at issue – when the interrogation *Q* is addressed to *C* and to *C'*. For th example to work its intended damage we would need to replace (1) by one single fixe question that computer *C* would answer correctly with YES and any other computer would answer correctly with NO. And this is impossible.

And one must also come to terms with the realization that no problem solver can ever reliably determine that all its more powerful compeers are unable to resolve a particular substantive problem (that is, one that is inherently tractable and not demonstrably unsolvable on logico-conceptual grounds). The plausibility argument for this is straightforward and roots in the limited capacity of a feebler intelligence to gain adequate insight into the operation of a stronger. After all, one of the most fundamental facts of epistemology is that a lesser capacity can never manage to comprehend fully the operations of a greater. The untrained mind cannot grasp the ways of the expert. And try as one will, one can never adequately translate Shakespeare into Pidgin English. Similarly, no problem-solver can determine the limits of what its more powerful compeers can accomplish. None can reliably resolve questions of computer solvability in general: none can reliably survey the entire range. The enhanced performance of a more capable intellectual performer will always seem mysterious and almost magical to a less capable compeer.

John von Neumann conjectured that computational complication is *reproductively* degenerative in the sense that computing machines can only *produce* others less complicated than themselves. The present thesis is that with regard to universal problem solving (UPS), computational complication is *epistemically* degenerative in that computing machines can only reliably *comprehend* others less complicated than themselves (where one computer "comprehends" another when it is in a position to tell just what this other can and cannot do).

Considerations along these lines substantiate that no computer in the field of general problem solving can obtain a secure overview of the performance of its compeers at large. And this means that:

Thesis T_1: *No computer can reliably determine that a given substantive problem is altogether computer irresolvable.*

Furthermore, it is clear that the question "But even if computers had no limits, how could a computer possibly manage to determine that this is the case?" plants an ineradicable shadow of doubt in our mind. For even if it were the case that computers had no problem-solving limits, and even if a computer could – as is virtually inconceivable – manage to determine that this is so, the fact would nevertheless remain that the computer could not really manage to secure our conviction with respect to this claim. Trusting though we might be, we would not – and could not reasonably be – *that* trusting. No computer could achieve credibility in staking so hyperbolic a claim. (We have the sort of situation that reminds one of the old Roman dictum: "I would not believe it even were it told to me by Cato.")

The upshot is that no computer problem-solver is in a position to settle the question of limits that affect its compeers across the board. We thus have it that:

Thesis T_2: *No computer can reliably determine that all substantive problems whatever are computer resolvable.*

But of course this thesis – like its predecessor – holds only in a domain which, like general problem solving (GPS), is totally open-ended. It does not hold for algorithmic decision theory (ADC) where one single problem solver (the Turing machine) can speak for all the rest.

The following dialectical stratagem also deserves notice. Suppose someone were minded to contest the acceptance of theses T_1 and T_2, and proposed to reject one of them. This very stance of theirs would constrain them to concede the other. For T_2 *follows from the denial of* T_1 (and correspondingly T_1 follows from the denial of T_2). In other words, there is no prospect of denying both these theses; at least one of them *must* be accepted.

The proof here runs as follows. If we let $X(p)$, as usual, represent $(\forall C)\sim C$ res P. Then we have:

T_1: For all C: $(\forall P)\sim C$ det $X(P)$

T_5: For all C: $\sim C$ det $(\forall P)\sim X(P)$

We thus have:

$\sim T_1$ = For some C: $(\exists P)$ C det $X(P)$

$\sim T_2$ = For some C: C det $(\forall P)\sim X(P)$ or equivalently C det$\sim(\exists P)X(P)$

Now the computers at issue are supposed to be truth-determinative, so that we stand committed to the idealization that C det P entails P. On this basis, $\sim T$ yields $(\exists P)X(P)$. And furthermore $\sim T_2$ yields $\sim(\exists P)X(P)$. Since these are logically incompatible, we have it that $\sim T_1$ and $\sim T_2$ are incompatible so that $\sim T$ entails T_2 (and consequently $\sim T_2$ entails T_1).

It is a clearly useful part of the plausibility-argumentation for T_1-T_2 to recognize that accepting at least one of them is inescapable.[23]

23 Principles T_1 and T_2 also have close analogues in general epistemology. T_1's analogue is $(\forall x)(\forall t)\sim KxUt$ or equivalently $(\forall t)UUt$ where $Up = \sim(\exists x)Kxp$. And T_2's analogue i $U\sim(\exists t)Ut$ which follows at once from the assertability of $(\exists t)Ut$.

References

Brennerman, H.J. 1977. "Complexity and Transcomputability," in Duncan, R. and Weston-Smith, M. eds. *The Encyclopedia of Ignorance*, (pp. 167–74). Oxford: Pergamon Press.

Davis, M. 1958. *Computability and Unsolvability*. New York: McGraw-Hill, (expanded reprint edition, New York: Dover, 1982).

Davis, M. ed. 1965. *The* Undecidable. Raven Press: New York.

Davis, M. and Hersh, R. 1978. "Hilbert's Tenth Problem," in Abbot, J.C. ed., *The Chauvenet Papers*, Vol. II, (pp. 554–571). Washington DC: The Mathematical Association of America.

Dreyfus, H.L. 1972. *What Computers Can't Do*. New York: Harper Collins.

Dreyfus, H.L. 1992. *What Computers Still Can't Do.* Cambridge, MA: The MIT Press.

Feferman, S. et al. 1995. eds. *Kurt Gödel, Collected Works, Vol. 3: Unpublished Essays and Lectures*. Oxford: Oxford University Press.

Hesken ed. 1988. *The Universal Turing Machine*. Oxford: Oxford University Press.

Leiber, T. 1997. "Chaos, Berechnungskomplexität und Physik: Neue Grenzen wissenschaftlicher Erkenntnis," in *Philosophia Naturalis*, 34, 23–54.

Penrose, R. 1989. *The Emperor's New Mind.* New York: Oxford University Press.

Pour-El, N.B. and Richards J.I. 1989. *Computability in Analysis and Physics*. Berlin: Springer Verlag. Hofstadter, D. 1979. *Gödel, Escher, Bach: An Eternal Golden Braid.* New York: Basic Books.

Rescher, N. 1998. *Predicting the Future*. Albany: SUNY Press.

Rogers, H. 1967. *Theory of Recursive Functions and Effective Computability.* New York: McGraw-Hill.

Shanker, S.G. ed. 1988. *Gödel's Theorems in Focus*. London: Croom Helm.

Weinberg, G.M. 1967. "Computing Machines," in Edwards, P. (ed.), *The Encyclopedia of Philosophy*, Vol. II, (pp. 168–173). New York: Macmillan & The Free Press.

An initial version of this paper was published in M. Carrier, G.J. Massey, and L. Ruetsche (eds.), *Science at the Century's End* (Pittsburgh, University of Pittsburgh Press, 2000), pp. 110–134.

PART 2

Two Poles of Analysis: Language and Ethics

Language and the Limits of Science

Ladislav Kvasz

Abstract

To see science as a human activity having some fundamental limitations is a Kantian insight. Even though his views have lost in many areas of philosophy their appeal, Kant's thesis about fundamental limitations of science is still compelling. If we try to base this thesis on language instead of on intuition, it is possible even today to argue in a technical way that every scientific theory has its *analytical* and *expressive boundaries*. By moving from one linguistic framework to the next one these boundaries can shift, nevertheless, they never disappear. In the present paper I would like to study the shifts of the analytical and expressive boundaries imposed on our thought by language. For this reason I will analyze several examples taken from the history of physics. One of the surprising outcomes of these analyses will be a new interpretation of Kant's antinomies of pure reason. If we relate Kant's antinomies not to reason as such, but to the linguistic framework of the particular physical theory, the antinomies retain their validity. In the form of the *analytical and expressive boundaries of language* the antinomies are a recurring feature of all physical theories. Thus Kant in his antinomies discovered the first manifestations of a universal epistemological fact.

Keywords

Language – limits – science – analytical – expressive – boundaries

The world of physics extends far beyond our sensory world. The properties of our everyday objects occupy only a tiny slice of the vast scales of physical quantities. Dimensions, temperatures, pressures, and densities which we know from our daily experience form only a small interval on the scale of dimensions, temperatures, pressures, and densities that appear in physics. Unlike the *physical world*, the *world of physics* changes along with how scientists proceed in uncovering ever more remote regions of the universe. In the world of physics there were no forces before Newton, there was no pressure before Torricelli, there was no entropy before Clausius. Of course, in the *physical world* forces,

 | DOI 10.1163/9789004325401_005

pressures, and entropy always existed[1] and their properties were independent from the development of our physical knowledge. In contrast to the physical world, the *world of physics*, i.e. the linguistic representation of the physical world by our theories, has undergone a number of fundamental changes. When Kuhn said that two physicists who support different paradigms live in different worlds, he had in mind different *worlds of physics* and not different *physical worlds*.[2]

The present article is a development of some ideas from my book *Patterns of Change* (Kvasz 2008). In the book I analyzed the linguistic innovations in the development of classical mathematics. In the present paper I shall turn to the main changes that occurred in the world of physics; changes of the way how we detect, study and represent physical phenomena, objects and processes. An illustration of such a change would be the transition from the Newtonian to the relativistic physics. In Newtonian physics, the world is described as a system of particles which move under the influence of forces acting at a distance in a three-dimensional empty space. In relativistic physics, the same physical world is described as a system of particles and fields which move, interact and radiate electromagnetic waves in a four-dimensional space-time.

The *most obvious* cases of this kind of changes in the history of physics were the creation of the mechanistic, the relativistic, and the quantum representations. Nevertheless, besides these most obvious cases there are several others, which will be shortly introduced in the first part of the paper.

1 An Overview of the Main Linguistic Frameworks in the History of Physics

In order to be able to introduce the notion of analytical and expressive boundaries of a physical theory in a rather precise technical way, it is important first to introduce two closely related but much less controversial notions, namely the analytical and the expressive power of a language (or of a linguistic

1 Even though it is not very clear what the terms *force*, *pressure* or *entropy* would mean were there no physics, let us suppose, for the sake of argument, that they have reference independent of our theories.

2 The problem, of course, is that scientists almost always and philosophers rather often *identify* the world of physics with the physical world. The scientists believe that their theories describe the *physical world*, while the (most of the analytic) philosophers of science think that there is nothing else in our reach than the *world of physics*. Thus they both ignore the fundamental difference between these two worlds.

framework). The *analytical power* of a linguistic framework can be characterized by the system of formulas which can be *derived analytically* by means of the particular framework. Thus by analytical power of the language of a physical theory I understand the system of all formulas (i.e. algebraic, differential or integral relations among the physical variables) which can be derived in the given language, using the accepted principles of the theory (i.e. without the use of additional empirical data or *ad hoc* hypotheses). As an illustration of the analytical power of the linguistic framework of Newtonian mechanic we can take Newton's derivation of Kepler's laws. For Kepler, the elliptical form of the planetary orbits was an empirical fact, i.e. a synthetic proposition. In the language of Newtonian mechanics it can be derived analytically from the law of universal gravitation. Thus, the possibility to make a derivation in a particular linguistic framework illustrates its analytical power.

The second important notion is the *expressive power* of a linguistic framework, which can be characterized as the ability of the framework to represent some aspect of the physical world. In the history of physics there are many cases when a phenomenon that defied description by means of the linguistic framework of some "old" theory and was thus seen as an anomaly could be unambiguously described by means of the linguistic framework of a "new" theory. For instance, heat conduction could not be described in the language of the Newtonian (particle) mechanics and required the introduction of the linguistic framework of the mechanics of continua and fluids in the 18th century, or that of statistical physics in the 19th century. Such cases can be interpreted as an increase of the expressive power of the language of physics.

It is interesting that besides the analytical and the expressive power of language, each linguistic framework can be characterized also by its analytical and expressive boundaries. By *analytical boundaries* of a particular linguistic framework I understand the fact that there are several laws, relations or facts, which can be expressed in the particular linguistic framework, but it is not possible to derive them within the framework. As a paradigmatic example of analytical boundaries of the linguistic framework of Newtonian physics we can take Newton's unsuccessful derivation of the speed of sound. It is clear that sound has a finite speed, Newton was able even to measure it with a reasonable accuracy, but the value he derived by means of his theory was incorrect. Nevertheless, the wrong result of Newton's derivation was not a failure of Newton himself, but rather a characteristic feature of the linguistic framework of his physics.

The second rather surprising kind of boundaries is the *expressive boundaries* of a linguistic framework. They concern relations or facts that cannot be even expressed by means of the particular linguistic framework. As an example we can take the relation $E = m.c^2$, which relates mass of an object with its total energy in the linguistic framework of the theory of relativity. Even though both energy and mass are concepts occurring already in Newtonian physics, in the Newtonian linguistic framework this relation is inexpressible, because there the speed of light is not a fundamental constant that could mediate such a relation. Only in the framework of the theory of relativity will the speed of light become a fundamental constant, and thus the above mentioned relation becomes expressible. In my opinion the expressive boundaries are one of the most interesting aspects of the language of science, and they shed new light on the nature of some of Kant's antinomies.

As the focus of the present conference is on the limits of science, I will give the analytical and the expressive boundaries of the language of science a systematic attention. But before turning to the analytical and expressive boundaries of the language of science, I will present a schematic overview of the different linguistic frameworks that were developed in the course of the history of physics (see figure 1).

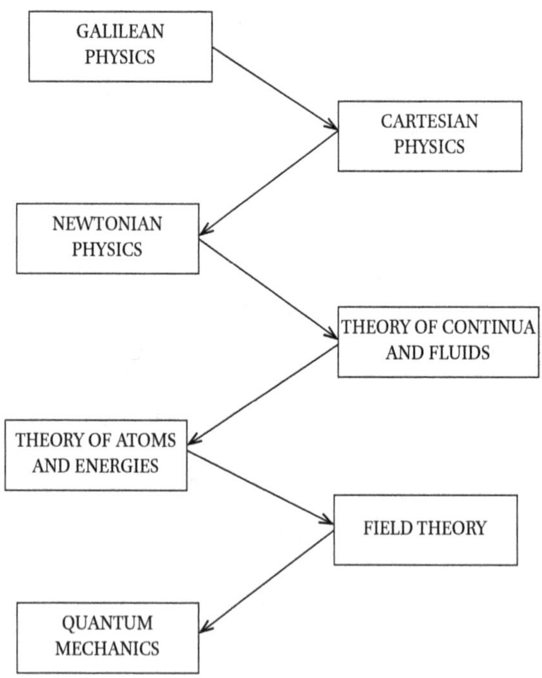

FIGURE 1 *Theories introducing new linguistic frameworks.*

Contemporary philosophy of science is devoted primarily to the discussion of three of these linguistic frameworks: the *Newtonian* framework, the *relativistic* framework (which is here subsumed under field theory), and the *quantum* framework. Some frameworks, as for instance the *Cartesian* framework or the fluidal part of the framework of *the theory of continua and fluids* are almost completely ignored, because the theories formulated on their basis were later discarded. Other frameworks are taken as provisory stages of some further development, and thus are not totally ignored, but their analysis is only short and superficial. Here it is perhaps not a proper place to argue that all seven above mentioned frameworks taken together constitute the historical development of the language of physics. This would require an independent and surely much longer paper. Instead I suggest taking the above list of frameworks as a working hypothesis (which is, of course, open to criticism) and focus on the limits these frameworks imposed on scientific research.

2 The Analytical and Expressive Boundaries of Language in the History of Physics

After an overview of the basic linguistic frameworks that occurred in the history of physics we can turn to the examination of their analytical and expressive boundaries.

2.1 *Galilean Physics*

Galilean kinematics is by many historians seen as the first modern physical theory. We will therefore begin our description of the boundaries of the language of physics by a brief analysis of the linguistic framework of this theory. The study of Galilean physics is, of course, a main theme in history of science. Galileo is perhaps best known for the discovery of the law of free fall, which in his own words is formulated as: "The spaces described by a body falling from rest with a uniformly accelerated motion are to each other as the squares of the time-intervals employed in traversing these distances" (Galilei 1638, p. 174).

This law represents perhaps the first scientific law – it is an experimentally established correlation between physical quantities. Galileo discovered several other similar laws, as for instance the law of the isochrony of the pendulum, the law describing projectile motion, or the law of motion on an inclined plane. A more detailed analysis of Galilean physics, compatible with the approach taken in the present paper can be found in (Kvasz 2002).

2.1.1 Analytical Boundaries of the Language of Galilean Physics

Despite his fundamental contributions, which are well known (see McMullin 1967 or Drake 1978), Galilean physics had also some grave shortcomings. Galileo seems to have had a too narrow concept of a natural law. All laws discovered by Galileo *lack generality*. Be it the law of free fall or the law of the pendulum, they are laws describing *particular phenomena*. In Galilean science, for each phenomenon there is a special law that describes it. It seems that these shortcomings have a common root. They are the consequence of the use of a too simple mathematics. Galileo believed that the book of nature is written in the language of mathematics:

"Philosophy is written in this grand book, the universe, which stands continually open to our gaze. But the book cannot be understood unless one first learns to comprehend the language and read the letters in which it is composed. It is written in the language of mathematics, and its characters are triangles, circles, and other geometric figures without which it is humanly impossible to understand a single word of it; without these, one wanders in a dark labyrinth" (Galilei 1623, pp. 237–238).

This passage is often quoted, but its strange nature is rarely acknowledged. Modern science is not based on any triangles, circles, or other geometric figures but on differential equations. The lack of generality, caused by this too simple mathematics, is a manifestation of the *analytical boundaries* of the language of Galilean physics.

2.1.2 Expressive Boundaries of the Language of Galilean Physics

Another peculiarity of Galilean physics is that Galileo's description of motion is always a description of the motion of a single, isolated body. The laws discovered by Galileo witness this. The law of the free fall, the law of the isochrony of the pendulum, or the law of the trajectory of projectile motion, these are all laws describing the motion of one isolated body. In Galilean physics *it is impossible to describe interactions between bodies*. The impossibility to describe interactions is a manifestation of the *expressive boundaries* of the language of Galilean physics.

2.2 *Cartesian Physics*

Cartesian physics developed a detailed (even though not always correct) theory of collisions, and so it can be seen as the first physical theory that was able to *describe interactions between bodies*. Thus Descartes was able to overcome the expressive boundaries of the language of Galilean physics. He described interactions by means of his law of conservation of the quantity of motion, which was perhaps the first *universal scientific law* in history. So, Cartesian physics was able to overcome also the analytical boundaries that hampered the development of Galilean science.

One of the main innovations of Cartesian physics was the introduction of its explanatory models. Descartes tried to explain physical phenomena by postulating mechanisms that caused them. For example he explained gravity by postulating a vortex of fine matter that pressed all bodies on the surface of the Earth downwards. This was an important step forward, because it indicates that Descartes clearly understood that the acceleration of the free fall must have some physical cause – it cannot happen just so.[3] I already mentioned that Galilean physics was unable to describe interaction. Against this background we can clearly see the conceptual progress introduced by Descartes. In Cartesian physics interaction in the form of the collision of bodies is a central phenomenon on which the representation of the world is grounded. Descartes described interaction by means of the law of conservation of the quantity of motion. For the simplest case of a totally inelastic central collision of two bodies it is thus possible to determine the resulting velocity.[4]

2.2.1 Analytical Boundaries of the Language of Cartesian Physics

There are so many errors in Cartesian physics that some historians completely excluded Descartes from their reconstruction of the history of physics. Many of these errors are, nevertheless, correctable within the Cartesian framework itself. However, besides these correctable errors (as for instance Descartes' incorrect rules describing collisions), there are problems that cannot be corrected within the Cartesian system. They rather reveal the boundaries of its linguistic framework. One of these fundamental problems is the identification of matter with space. Due to this identification Descartes has no independent reference system on the foil of which he could define the direction of a motion or the shape of a trajectory. Descartes defines motion purely relatively, as "the transference of one part of matter or of one body from the neighborhood of those bodies that immediately touch it and are regarded as being at rest, and into the neighborhood of others" (Descartes 1644, part II sec. 25).

In such a framework *it is impossible to define rectilinear motion.* How can we detect rectilinear motion in the case when all surrounding bodies move? Nevertheless, the concept of a uniform rectilinear motion is one of the fundamental concepts of the whole Cartesian physics and it plays a crucial role in

3 Galileo totally missed this point.

4 For a detailed discussion of Cartesian physics see Shea (1978), Gabbey (1980), Gaukroger (1980), Cottingham (1992), Shea (1991), or Kvasz (2003).

the formulation of Descartes' second law of nature. This inconsistency of the Cartesian system was sharply criticized by Newton in *De Gravitatione* (Newton 1670). And we cannot change Descartes' definition of motion without a radical reconstruction of the entire system.[5] The impossibility to define rectilinear motion illustrates the *analytical boundaries* of the language of Cartesian physics.

2.2.2 Expressive Boundaries of the Language of Cartesian Physics

Another fundamental flaw of the Cartesian system, the clarification of which was the purpose of the second book of Newton (1687), concerns friction. Newton showed that if gravity were caused by a vortex of fine matter, as Descartes maintained, then within a short period of time the whole mechanism of the solar system would come to a halt. In order to mediate the exchanges of the enormous momenta that accompany the motion of celestial bodies (during the period of six months the Earth changes the direction of its motion to the opposite one, which, given its weight and its speed, means an enormous, almost unimaginable, change of momentum) the vortex of fine matter must interact very intensively with matter. Therefore gravity cannot be caused by a vortex of some rarified ether. The vortex must consist of the motion of a dense fluid, capable of transferring large momenta. If so, then any rapid motion through such a fluid would be accompanied by considerable friction, which would bring the motion of the Earth to a halt. Nevertheless, the fact that Descartes did not incorporate friction into his system is not a coincidence. It is simply impossible to incorporate friction into a system that describes interaction by means of conservation laws. Friction violates the law of conservation of the quantity of motion, because due to friction a portion of the quantity of motion is lost (in the form of heat).[6] Since Descartes described interaction by means of the law of conservation of the quantity of motion, *he could not describe friction*, because it contradicts this law. Therefore Newton's criticism revealed not an error, but rather the expressive boundaries of the linguistic framework of Cartesian physics.

---

5 Newtonian physics can be, at least partially, seen as an outcome of such a reconstruction.

6 It is important to remember that the Cartesian quantity of motion is a scalar quantity, it is the sum of the products of the velocity and magnitude of each body, the velocity understood as a scalar quantity. Thus a decrease of the quantity of motion cannot be compensated (as in the case of the Newtonian momentum) by a similar decrease of the same quantity in the opposite direction. In the Cartesian system, where friction decreases velocity, a particular amount of this quantity is lost, and thus the conservation law is violated.

2.3 *Newtonian Physics*

Newton described interaction by means of forces acting at a distance. For a body to be able to act on another body, it is not necessary, according to Newton, that they are in a direct contact. Bodies can act on each other even when they are apart. An example of such an action at a distance is the Newtonian force of gravitation. For Descartes interaction consisted in collision accompanied by a passive transference of momentum from one body to another. In contrast to this, Newtonian interaction is an active process. In the Newtonian system *forces can generate motion*. Therefore, the total amount of motion (if we understand it like Descartes did, as a scalar quantity mv) is not preserved. When a body falls to the Earth, new quantity of motion is generated. According to Newton, the impulses of a force accelerate the body and so generate motion.[7]

One of the great triumphs of Newtonian physics was the *derivation of Kepler's laws from the law of universal gravitation*. Kepler's law about the elliptical form of the planetary orbits was an empirical proposition. Kepler discovered it by analyzing the data about the motion of Mars. In Newtonian mechanics, this law can be derived from the law of universal gravitation. Of course, historically it went the other way round – Newton used Kepler's laws in his derivation of the law of universal gravitation. But the direction in which we move from one law to another is irrelevant. What is important is the existence of analytical ties between these two propositions, and these ties illustrate the analytical power of the language of Newtonian physics. In the short description of the Cartesian system I stated that its language is not able to describe friction. For the Newtonian physics the description of friction presented no problem. Newton described interactions by means of his second law. If there was friction in the system, it just meant that in the particular equation Newton had to introduce a further term, corresponding to the force of friction.

---

7 In Newtonian mechanics *momentum* is conserved instead of the Cartesian *quantity of motion*. The momentum is a vector quantity, while the quantity of motion is a scalar one. However, there is a further difference between them. The Cartesian quantity of motion is a substance; according to Descartes, there is just as much motion in nature as God put there during creation. The momentum, on the other hand, is not a substance. Its conservation is not the consequence of someone putting at the beginning a certain amount of it into the universe and later neither adding nor removing any part of it. The conservation of momentum is a consequence of the law of action and reaction, because according to this law forces simultaneously generate equal amounts of momentum in opposite directions, so that the total momentum remains unchanged. In the case of the free fall of a body, the Earth is also accelerated, not just the body, and so the total momentum of the system containing the body *and the Earth* remains constant.

2.3.1 Analytical Boundaries of the Language of Newtonian Physics

Illustrations of the analytical boundaries of the language of Newtonian physics are the problems where Newton calculated a result that contradicted the experimental data. Let us consider the *calculation of the speed of sound* (Prop. L, Prob. XII of the second book of the *Principia*), where he derived 968 feet per second, i.e. 295 ms^{-1}, which is 17% less than the actual value of the speed of sound. Newton's own experiments at Trinity College gave a value well above this theoretical prediction, and so he embarked on a series of emendations of his derivation, which Westfall called an "*egregious fraud*" (Westfall 1971, p. 497). Similarly problematic is Newton's calculation of time in which a fluid escapes through a hole in the bottom of the container (Prop. XXXVI, Prob. VIII of the second book, of the second edition of *Principia*). Here he obtained twice the value that he got in the measurement. Considering this problem Westfall notes that here "*the dynamics completely failed*" (Westfall 1971, p. 501).

In these examples, apart from the purely mechanical side of the problem, which Newton correctly understood, there was also an additional aspect (a thermodynamic or a hydrodynamic one, respectively), which Newton did not take into account, and which was responsible for the considerable deviation from the experimental results. Nevertheless, a full understanding of these additional aspects was not possible until the transition from the Newtonian physics to the theory of continua and fluids (discussed in the next chapter). Therefore I think that these failures do not constitute an "*egregious fraud*" or a "*complete failure*" of Newton's dynamics. They can be interpreted as manifestations of the analytical boundaries of its language. It is not surprising that the above mentioned derivations failed. In the language of Newtonian physics the speed of sound or the time of leakage cannot be derived.[8] The remarkable fact is that, despite the inappropriateness of the language of Newtonian physics for the description of thermal and hydrodynamic phenomena, Newton got a roughly correct result. This shows his brilliance.

8 I use the term "Newtonian physics" in a narrower sense than it is used in the physical community. From the perspective of the present paper it is important to realize that Newton did not use partial differential equations. Physicists consider the introduction of this kind of equations a "technical detail" and so downplay the conceptual changes involved. Here we can quote Newton from a letter to Halley: "*Mathematicians that find out, settle, and do all the business must content themselves with being nothing but dry calculators and drudges and another that does nothing but pretend and grasp at all things must carry away all the invention as well of those that were to follow him as of those that went before,*" Westfall (1980, p. 448). Here Newton had Hooke in mind. Hooke made a claim to the discovery of the law of universal gravitation. His words, however, have a deeper meaning and indicate that Newton did not consider the mathematical aspect of physics in any sense a subordinate. Thus we should neither.

2.3.2 Expressive Boundaries of the Language of Newtonian Physics

Newtonian physics was able answer many questions about the dynamics of the solar system. Nevertheless, this theory is far from being complete. One of the first who realized this was Immanuel Kant (in Kant 1781), who formulated a series of antinomies revealing the boundaries of the Newtonian description of the world. In Kant's times, Newtonian physics was the only physical theory, and so Kant considered its antinomies to be properties of reason as such. He did not attach them to a particular physical theory, but he saw in them boundaries that limit our very ability to create any theory. If this were true, the antinomies would be insurmountable.

However, as the development of physics showed, Kant's antinomies can be overcome. For example, the general theory of relativity eliminated the antinomy of the finiteness versus the infiniteness of space, when it replaced the Euclidean space by the curved space-time, for which Kant's antinomy does not work. In spite of this, Kant's antinomies do not lose their importance. The only thing we have to do is to relate them to the particular language in which the theory for which Kant asserted the antinomy was formulated. Thus we propose to interpret the antinomy of the finiteness versus the infiniteness of space, as a reflection of the *external nature of space in Newtonian physics*. In our view, this antinomy is not a feature of human reason but rather of language. And it is a feature not of language as such, but of the language of Newtonian physics. The antinomy of the finiteness versus the infiniteness of space can be interpreted as a manifestation of the *expressive boundaries* of the language of Newtonian physics. It indicates that in the linguistic framework of Newtonian physics it is not possible to express the relationship between space and matter (the relationship the description of which is the core of the general theory of relativity).[9]

---

9 When we declare the antinomy of finiteness or infiniteness of space to be a manifestation of the expressive boundaries of the language of Newtonian physics, which were overcome by the theory of relativity, we don't claim that relativistic physics can give a definitive answer to this problem. We are claiming only that for the theory of relativity the question about the finiteness of space is no longer a speculative question. When we view as the source of Kant's antinomies not human reason, but rather the language of a particular physical theory, we can interpret Mach's criticism of the notions of absolute space and absolute time (Mach 1893) as a deepening and radicalization of the Kantian position. Mach discovered the external character of space in Newtonian physics. He thus diverted the sting of Kant's criticism from the limits of pure reason to the limits of a particular physical theory. Our approach can then be seen as a further development of Mach's criticism. We claim that problems analoguous to those discussed by Mach are not restricted to Newtonian physics, but that they are a systematic feature of the language of any physical theory. Thus, this aspect of science is universal,

2.4 *The Theory of Continua and Fluids*

In the Newtonian universe interaction is mediated by forces acting at a distance. This finds its expression in the fact that the equations of motion have the form of ordinary differential equations. In 1713, when trying to describe the vibrating string, Brook Taylor hypothetically distinguished an element of the string and examined the forces acting on it from the side of the contiguous elements. In principle he only applied to the element of the string Newton's second law, and therefore believed that he proceeded in the spirit of Newtonian orthodoxy. A little later, in 1736, Euler formulated a research program that systematically studied forces acting inside of substance (Euler 1736). The first trace of awareness of the fact that this program abandoned the framework of Newtonian physics is the title of Euler's work: *Discovery of a new principle of mechanics* from 1750, where Euler formulated the principle according to which the differential equations describing the motion of a free body remain valid also if they are used to describe not the body as a whole (as Newton did), but for the description of an element of a body or a fluid.

During the 18th century the theory of continua and fluids was created. It represents an extended body as if composed of parts – the elements of the continuum – , which are *hypothetically distinguished*. These elements have the same characteristics as the continuum as a whole (the same density, elasticity, hardness) but are so small as to allow the transition to differentials. In mathematical terms this meant that besides the ordinary differential equations of Newtonian mechanics describing the motion of the body as a whole, a new kind of equations emerged – the partial differential equations. These are equations such as the equation of the *vibrating string*, the equations of *fluid dynamics* and the equation of *heat conduction*. They describe the motion of a continuum or the spread of activity (stress, compression, etc.) in a continuous medium. For the description of some phenomena, new fluids were postulated (the electric fluid, the magnetic fluid, the caloric, the phlogiston, the ether). These theories resemble Cartesian physics, and many historians see in Euler a successor of the Cartesian program, or even a hidden Cartesian. In my view this is a mistake. Euler described the action of forces by means of differential equations, just like Newton. Thus he did not make a step backwards in the direction of Descartes, but went forward towards a new linguistic framework.

The linguistic framework of the theory of continua and fluids made it possible to derive differential equations describing different physical processes

---

just like Kant believed. The only difference is that it is not related to reason but the language of physical theories.

such as vibrations of a string, flow of water, heat conduction or electric currents. Perhaps the most spectacular achievements were Fourier's derivation of the equation of heat conduction published in 1822 in his *Théorie analytique de la chaleur*, and Carnot's derivation of the formula describing the effectiveness of thermal machines published in 1824 in the *Réflexioin sur la puissance motrice de feu*. Fourier derived his equation and Carnot proved his formula under the supposition that there is a fluid that they called *caloric*.

One of the few mistakes in Newton's *Principia* was the derivation of the speed of sound. Newton calculated it by means of a clever trick, where he likened the vibrations of air in the sound wave to a mechanical pendulum (the necessity to use such an ingenious trick indicates that the description of sound waves escapes the possibilities of the language of Newtonian mechanics). The correct value of the speed of sound was derived by (Laplace 1816), when he realized that the compression of air in the sound wave is not isothermal, as Newton implicitly assumed. During the phase of compression, the air in the sound wave gets heated and this increase of temperature raises the speed of the sound waves. A theoretical justification of Laplace's derivation was given in 1823 by Poisson, who assumed that the amount of heat contained in a given volume of air remains constant during its compression. Such processes are called *adiabatic*. The notion of an adiabatic process belongs to the theory of continua and fluids. If we imagine the air as a sponge imbued with *caloric*, then an *isothermal compression* of the air corresponds to a slow pressing of the sponge by means of which some *caloric is squeezed out of the sponge*. The compressed sponge occupies a smaller volume and this smaller volume can take in only a smaller amount of *caloric*. The superfluous *caloric* leaves the sponge. When the pressing of the sponge is too fast, and in the sound wave that we are dealing with here compression and expansion alternates 1000 times per second, the *caloric* does not manage to leave and it is pressed together with the sponge. The pressing of the *caloric*, i.e. the increasing of the amount of *caloric* in a unit of volume, is nothing else but a rise of temperature. Poisson realized that in sound waves the vibrations run so fast that the *caloric* is compressed together with the air, and so the condition of an isothermal process is violated.

Here we see how the idea of heat as a fluid makes it possible to analyze the thermal conditions in the sound wave, to *express the difference between an isothermal and an adiabatic process* and thus to arrive at the correct value for the speed of sound waves. For technical applications similarly important is the description of the process of deformation of a body under a load or at impact. While Newtonian physics described the process of deformation of a body at impact as a whole, by means of the total force of elasticity, the theory of continua and fluids can describe the process of propagation of the deformation in the body and makes it possible to calculate the stress at different points of the body.

2.4.1 Analytical Boundaries of the Language of the Theory of Continua and Fluids

The theory of continua and fluids describes the thermal phenomena by postulating the existence of a new substance, the *caloric*. Then it describes heat conduction as the flow of caloric through the pores of matter, and the increase of temperature as an accumulation of caloric in a given volume. In 1843 Joule determined the mechanical equivalent of heat, thus showing that mechanical work can be converted into heat. The emergence of heat from mechanical work contradicts the idea that heat is a substance (a fluid). Joule's experiments thus show the analytical boundaries of the language of the theory of continua and fluids – this language cannot derive the generation of heat by mechanical work.

2.4.2 Expressive Boundaries of the Language of the Theory of Continua and Fluids

Just as Kant's antinomy of finite versus infinite space illustrated the expressive boundaries of Newtonian physics, the antinomy of finite versus infinite divisibility of matter can be taken as an illustration of the expressive boundaries of the language of the theory of continua and fluids. Kant's antinomy shows that the notion of a volume element in the theory of continua and fluids is external. The divisibility of matter is postulated here (just like in Newtonian physics the infinite space was postulated) but it is not empirically fixed. When in the theory of continua and fluids we discriminate a volume element in a continuum, it is a mathematical operation not backed by any experimental procedure that could determine the physical properties of matter at these scales of magnitude. We simply assume that matter is homogeneous and so the element of volume has the same properties as the macroscopic body as a whole. Kant pointed here to a fundamental problem, but as in the previous case, I do not ascribe this antinomy to reason as such. Quantum mechanics removed this antinomy when it showed that due to the uncertainty principle any further division of matter is accompanied by an increase in energy. The *speculative* question of divisibility of matter became a practical question of availability of ever higher and higher energies, which ultimately lead to a technical problem of building larger and larger accelerators. This shows that Kant's antinomy belongs not to reason but that it outlines the expressive boundaries of the language of the theory of continua and fluids.

2.5 *The Theory of Atoms and Energies*

The theory of atoms and energies emerged from a crisis of the mechanistic worldview, which was the consequence of the gradual progress in several scientific disciplines in the second half of the 18th century. In chemistry it was discovered that air is not a simple elastic continuum, as it was represented by the theory of continua and fluids, but it is a mixture of several different substances. In 1755 Black discovered the carbon dioxide (*fixed air*), in 1766 Cavendish discovered the hydrogen (*inflammable air*). These discoveries culminated in 1789 in Lavoisier's oxidation theory of combustion, which replaced the *phlogiston* theory and led to the creation of the notion of a chemical element. Development of calorimeters led to Joule's measurement of the mechanical equivalent of heat, which discredited the notion of *caloric*. Thus Lavoisier discredited the phlogiston, Joule the caloric and finally the theory of relativity discredited the last weightless fluid – the *eather*. Advances in the theory of materials such as optical glass or steel during the first half of the 19th century brought physics beyond the limits of the theory of continua and fluids. These advances fundamentally changed our understanding of the structure of matter.

Around the middle of the 19th century a transition occurred *from the hypothetical postulation of mathematical continua or weightless fluids to an empirical study of the structure of materials and of the processes taking place in them*. The emergence of a new linguistic framework is closely linked to the abandonment of fluids. This new framework abandoned the mathematically postulated fluids on which the previous linguistic framework was based, but tried to maintain the results that were achieved by it. Physics moved one level deeper in the description of the structure of matter. The theory of continua and fluids simply postulated continuous substances representing macroscopic phenomena like fire, heat, and electricity, and created a mathematical language which made it possible to calculate their behavior. Thus the derivation of Fourier's equation for heat conduction was based on the assumption of the existence of *caloric*, just like the derivation of Maxwell's equations of electrodynamics was based on the assumption of the existence of *aether*. With the increase in the accuracy of experimental methods, physics was able to take a step beyond the macroscopic level. The macroscopic properties that the theory of continua and fluids tried to explain by means of postulating *hypothetical substances* became statistical averages of properties of *real particles* constituting the microscopic level of description.

As in the cases of the previous languages, also in the case of the language of the theory of atoms and energies, the first hints of what would later become a new linguistic framework emerged as technical tricks designed to solve particular problems for which the standard methods of the old language were not suitable. Henri Navier, who contributed to the development of the theory of continua and fluids by a definition of the modulus of elasticity and by the experimental determination of its value for iron, submitted in 1822 his *Mémoire sur les lois du mouvements des fluides*, where he derived the *equation of motion of an incompressible viscous fluid*. Although the equation itself contains only variables characterizing the fluid as a continuum, Navier derived it on the assumption that the liquid consists of molecules, the forces of interaction between which are proportional to their mutual velocities. Thus the notion of a molecule occurred in the theory of continua and fluids as a trick, allowing to derive the equation of motion of the fluid (just like Taylor introduced material forces a century earlier and just like Planck's quanta will occur some eighty years later). Navier's idea does not fit into the "orthodox" theory of continua and fluids: a molecule is something fundamentally different from an element into which the continuum ought to be cut according this theory. The macroscopic properties of the continuum are not simply transferred onto the molecules, as the theory of continua used to transfer the macroscopic properties onto the elements of the continuum. On the contrary, the molecules have properties that are different from the macroscopic properties of matter, and the properties of the continuum are derived statistically (and not simply transferred) from the properties of the molecules. Atoms or molecules, from which a liquid is composed, are *not hypothetical entities* postulated mathematically, but *physically real objects*, although their size, number and characteristics were little known. *Physically real* in this context means that the discrimination of atoms or molecules is not a hypothetical act of definition (as in the case of the volume element of a continuum), but that atoms are taken to be real, empirically detectable material particles.

Perhaps the best illustration of the expressive power of the language of the theory of atoms and energies is chemistry. In early 19th century chemistry had a long history of successful empirical research, but only the theory of atoms and energies was able to put the results achieved in this research on solid theoretical foundations.

2.5.1 Analytical Boundaries of the Language of the Theory of Atoms and Energies

If we consider matter as composed of atoms and atoms as tiny balls, there is no reason why these tiny balls could not rotate and vibrate. Assuming that the laws of mechanics apply on the atomic level, each atom becomes a rigid body with an infinite number of degrees of freedom, one for every mode of internal oscillations. In the state of thermal equilibrium at temperature T each degree of freedom has energy equal to kT. Since the energy of oscillations is proportional to their amplitude, internal oscillations can be excited with arbitrarily small energy. From the fact that there is an infinite number of internal degrees of freedom which can absorb arbitrarily small portions of energy it follows that *eventually all energy will be absorbed by internal degrees of freedom of atoms*, which contradicts our experience. It might be objected that also every macroscopic body has an infinite number of internal degrees of freedom and so this paradox had to appear already in the theory of continua and fluids. But this is not true. The theory of continua and fluids understood heat as a fluid, so the thermal equilibrium and the oscillations of the body were unrelated.[10] Only when the theory of atoms and energies interpreted heat as the energy of atomic motion, a relation between the internal degrees of freedom and the distribution of heat could appear.

2.5.2 Expressive Boundaries of the Language of the Theory of Atoms and Energies

In 1820 Andre Marie Ampère discovered the force acting between two electric currents. Loyal to the principles of the theory of continua and fluids, he formulated a quantitative law describing how two elements of current act on each other. When in 1897 it turned out that an electric current represents the collective motion of charged particles, from Ampère's law it followed that the electric current acts also on the moving charged particles. If we confine ourselves to a single particle, from Ampère's law we can calculate the force that acts on it from the side of the current. But here lurks a problem. If we look on the same situation from the viewpoint of the coordinate system joined to the charged particle, in this system the particle is motionless and the current moves. Nevertheless, an electric current does not exert any force on a motionless particle. Thus, whether a force acts on a charged particle from the side of a current or not, depends on the choice of the coordinate system. But forces exist independently of our choice of coordinates. This shows that in the theory of atoms and energies *the change of the coordinate system has an external character* and is not tied to experimental procedures.[11]

10 It was precisely this unrelatedness of thermal and mechanical phenomena that illustrated the analytical boundaries of the linguistic framework of the theory of continua and fluids.

11 This paradox may be added to the list of Kant's antinomies, as it has a similar formal structure – the external character of a particular object. It indicates the expressive boundaries of the language of the theory of atoms and energies. We can formulate a thesis – an electric current acts on an electric charge; and an antithesis – an electric current does

2.6 *Field Theory*

The notion of an electromagnetic field emerged from the experimental work of Michael Faraday, who introduced the concept of field lines to visualize the action of electric and magnetic forces on charges, currents and magnets. Faraday used this concept in the description of the electromagnetic induction he discovered in 1831. Most physicists did not take Faraday's lines of force seriously, seeing in them only a heuristic device that may be helpful in the discovery of new facts, but adds nothing to the physical content of the theory. That Faraday's field lines are not only a heuristic device enabling us to visualize the processes involved in an experiment, but that they also have physical content, was realized by James Clark Maxwell, who in (Maxwell 1861) rewrote Faraday's qualitative considerations into a mathematical form and gradually turned Faraday's concept of field lines into that of an electromagnetic field. Maxwell's ideas were further developed by Hendrik Lorentz who incorporated into Maxwell's theory a description of the interaction between the field and matter. Lorentz wanted to reconcile Maxwell's field theory with the theory of atoms and energies. To this end he introduced laws describing the contraction of atoms in motion, now called the *Lorentz transformations*. Gradually it became clear that no such reconciliation is possible and in 1905 independently Albert Einstein and Henri Poincaré concluded that field theory is a fundamentally new linguistic framework, requiring a new interpretation of the categories of time and space. The speed of light, which formerly characterized only the spreading of light and of electromagnetic waves, i.e. a relatively limited range of phenomena, suddenly appeared in the definition of mass or in the equations describing the transformation of coordinates. Field theory was no longer a theory describing a limited class of phenomena, as Maxwell understood it. It became a linguistic framework used in the description of the entire world of physics.

The analytical power of the language of field theory can be illustrated by Maxwell's discovery of the displacement current. When he rewrote all the known facts about electric and magnetic fields into a mathematical form, Maxwell found out that the equations he received are asymmetrical. A changing magnetic field generates an electric field (Faraday's law of electromagnetic induction), but a changing electric field had no analogous effect. Led by the idea of symmetry, Maxwell postulated the existence of a magnetic field generated by a changing electric field. This effect was not yet discovered because its detection requires very special conditions which cannot be hit upon by chance.

---

not act on an electric charge; and just like Kant did in the other cases, show that both are untenable.

When he supplemented the equations by an additional term, he found that they have a solution in the form of electromagnetic waves. Maxwell published his discovery in 1873 and 1886 Heinrich Hertz proved experimentally the existence of electromagnetic waves. Shortly thereafter a hitherto unseen technical development started from the telegraph and radio through television and radar to telecommunications satellites and cell phones. Nevertheless, it is important to keep in mind that all these developments started on a sheet of paper. The transcription of experimental data into a mathematical form revealed a gap in the experimental results and the filling of this gap by a new term in the equations led to the prediction of electromagnetic waves. When Maxwell calculated the velocity of these waves, he obtained the speed of light. This led him to the idea to interpret light as electromagnetic waves. Maxwell was thus able to *derive the laws of optics from the laws of electrodynamics*. This derivation can be seen as an example illustrating the analytical power of the language of field theory.

We illustrated the analytical boundaries of the language of the theory of atoms and energies by the following paradox: When we turn in our description of a moving charged particle to the coordinate system coupled to that particle, the force by which an electric current acts on this particle disappears. In the framework of field theory this paradox can be explained. The trick is that from the point of view of the coordinate system coupled to the flying particle, the conductor in which the electric current flows is moving. As a result of this motion a relativistic length contraction in the moving conductor will occur. This contraction will shorten the intervals between the positive charges of the metal grid as well as those between the negative charges forming an electric current. Nevertheless, as the grid and the negative charges (forming the electric current in the grid) move with different velocities, the corresponding contractions of the intervals between the positive charges of the grid and the negative charges constituting the current will be different. This difference will lead to the formation of a non-compensated charge on the conductor. Thus, the conductor ceases to be electrically neutral, and so our charged particle will be subject to a force also in the coordinate system coupled to it. This force will produce the same effects as Ampère's law predicted in the original system. Thus the paradox is successfully removed.

2.6.1 Analytical Boundaries of the Language of Field Theory

During the 19th century the experimental research of black body radiation achieved considerable results. Physicists measured the curves that indicate the intensity of radiation at different lines of the spectrum for many different temperatures. Wilhelm Wien formulated in 1894 a law that approximated these curves rather well at high frequencies, but at low frequencies it led to divergence (the so called infrared divergence). In 1900 John Rayleygh and James Jeans derived another law which accurately described the curves at low frequencies, but led to a divergence at high frequencies (the so called ultraviolet divergence). Thus there were laws working at both ends of the spectrum, but it was not possible to connect these asymptotic laws. The *black body radiation* is thus a phenomenon that characterizes the analytical boundaries of the language of field theory. The linguistic framework allows us to derive formulas which for different segments of the spectrum agree nicely with the data, but the agreement is only partial, just like in the case of Newton's derivation of the speed of sound.

2.6.2 Expressive Boundaries of the Language of Field Theory

Field theory cannot explain how is it possible that the bodies around us are stable and do not alter their shape. Electric and magnetic forces have an interesting feature – for fundamental reasons they cannot sustain a *stable configuration of charged particles*. The reason is as follows: Suppose that we would like to create a configuration of several charged particles so that it would be stable with relation to small perturbations. This would mean that if we choose one of these particles, the other particles should create in its vicinity such a field that after a small change of the position of our chosen particle from its stable position, the forces of the field would return it back (this is the meaning of the notion of stability). Thus if we imagine a small sphere around our chosen particle, so small that there are no other charges in it, the lines of the field generated by the remaining particles must intersect this sphere pointing inwards (so that if the chosen particle tried to leave the sphere the field would return it back). But according to Maxwell's equations this is not possible.

Trying to create a stable position as a dynamic configuration cannot save the situation. Everyone knows the toy – the whipping top – that is unable to stay at its top, but when you spin it, it easily does so. Therefore, one could think that in a similar fashion even if a static configuration of charges cannot be stable, a stable configuration can be constructed as a dynamic configuration of charged particles. Unfortunately, this hope is quickly shattered by Maxwell's equations, because in such a case the motion of the charged particles must be along curved paths. But by moving on curved paths, the charged particles would emit electromagnetic radiation and thus constantly lose energy. Thus in the case of an atom the electrons would after a very short time fall on the nucleus. Matter must therefore be held together by something that is inexpressible in the language of field theory. For field theory, the stability of matter is a mystery.

2.7 *Quantum Mechanics*

The paper in which the quantum hypothesis was introduced for the first time was written by Max Planck; it appeared in 1900 and concerned black body radiation. As we stated above, the attempts to describe black body radiation in the framework of field theory led to divergent formulas. Planck was able to reach a satisfactory outcome – the well known Planck's formula – but at the price of the hypothesis that the black body does not emit its radiation continuously, as required by the principles of classical physics, but in small discrete portions, which he called *quanta*. Planck considered the quantum hypothesis a "dirty trick", and he hoped to find an alternative derivation of the radiation law that would not contradict the principles of classical physics. In 1905 Einstein used Planck's hypothesis in his theory of the photoelectric effect and Bohr incorporated it in 1913 into his theory of atoms. While Planck understood the quantum hypothesis as a trick (he believed that it will be possible to find a derivation of the radiation law without its use), Bohr and Einstein ascribed ontological reality to quanta. They believed that besides atoms and electrons there is another kind of objects: the quanta of radiation.

In 1923 Louis de Broglie came to the conclusion that the quantum hypothesis concerns not only radiation. In a similar way as Plank assigned to the continuous radiation his discrete quanta, it is possible to assign to discrete particles some continuous waves. These material waves were not yet detected, because their wavelength is extremely small. Thus quanta were no longer a special kind of objects, as they were understood from 1905 till 1923. The quantum hypothesis was turned into a universal principle valid for all objects. It became a basis for a new linguistic framework. In this framework, all physical systems manifest wave-particle dualism. Thermal radiation was the area where physicists for the first time came across this principle. After the work of de Broglie in quick succession followed the works of Heisenberg, Born, Jordan, Schrödinger, Dirac, Pauli, and in 1927 von Neumann created the standard mathematical formulation of quantum mechanics, based on the notion of Hilbert spaces.

The analytical power of the language of quantum mechanics can be illustrated by the derivation of *Planck's radiation law*. This formula, which is in good agreement with observation, cannot be derived without the quantum hypothesis. The divergence of Wien's law and of Rayleygh-Jeans' law is not a result of mathematical incompetence of their authors. In the linguistic framework of classical physics it is not possible to derive the correct formula for black body radiation. Another great success of quantum mechanics is the description of the electron shells of atoms and of the related phenomena such as atomic spectra and chemical reactions. For classical physics the existence of spectra that are characteristic for each chemical element or compound was a mystery. Quantum mechanics allows us to calculate the spectra of atoms with few electrons, and to approximate with sufficient accuracy the spectra of more complicated systems. Similarly, quantum mechanics succeeded to calculate the binding energies of the simplest molecules and for the more complex molecules it developed a framework in which it is possible to describe chemical reactions.

In connection with the analytical boundaries of the language of field theory we stated that this language cannot represent a stable system of charged particles. In quantum mechanics there is the so-called *Heisenberg's uncertainty principle*, which allows just that. The uncertainty principle says that the product of the uncertainty of coordinates Δx and the uncertainty of momentum Δp must be greater than Planck's constant h. When we imagine a system consisting of two particles, a positively charged proton and a negatively charged electron, then from the point of view of field theory they cannot create a stable spatial configuration, because an electron circling round a proton would radiate electromagnetic radiation, thereby gradually losing energy and would therefore fall on the proton. The system should, according to field theory, collapse. Therefore a stable hydrogen atom should not exist. But here the uncertainty principle becomes relevant. It prevents the electron from coming too close to the proton, because then its coordinate would be very well located (in atoms the radius of the atomic core is hundred thousand times smaller than the radius of the atom itself, therefore a fall on the core means an increase of the accuracy of the electron's localization by five orders of magnitude). This increase of accuracy of localization in space causes due to the uncertainty principle a great increase of uncertainty of the value of momentum. An increase in accuracy of the position by five orders of magnitude would lead to a decrease of accuracy of the momentum by the same five orders of magnitude. But this huge uncertainty of the momentum means that the electron would leave its "trajectory" along which it is to fall on the proton according to the classical theory. Thus, Heisenberg's uncertainty principle acts against the collapse of the system. Electric attraction tends to push the system into the smallest possible area in space, while the uncertainty principle pushes it out from these areas. The stable ground state of atom emerges as a compromise between these two tendencies.

2.7.1 and 2.7.2 Analytical and Expressive Boundaries of the Language of Quantum Mechanics

I will not introduce the analytical and expressive boundaries of the language of quantum mechanics. The critique of the Copenhagen interpretation of quantum mechanics from different positions is well known, but it is unclear whether the dissatisfaction with the foundations of quantum mechanics is only the result of its unusual conceptual framework or it is a real weakness of this theory. In the cases discussed above we characterized the analytical and expressive boundaries of a linguistic framework by means of its confrontation with some later language, which transcended these boundaries and made it thus possible to characterize them. In the case of quantum mechanics we can consider only quantum field theory in the role of the later language, the formalism and conceptual foundations of which seem too complicated for a non-technical philosophical reflection.

3 Summary

In contemporary philosophy of science it is widely held that the only limits a theory may encounter are empirical in nature. Our reconstructions showed that each physical theory has important internal limitations, the overcoming of which is (besides the empirical progress) one of the main driving forces of the development of physics. It seems that in the form of his famous antinomies Kant discovered that scientific theories have fundamental limitations. Nevertheless, he did not bring these limitations into contact with the language of science, but tried to construe them as a consequence of the structure of human reason. When we interpret Kant's theory of antinomies linguistically, it becomes possible to separate the epistemological core of Kant's discovery from its contingent formulation. Thus, our reconstruction of the history of physics can bring a new impetus for the understanding of Kant's philosophy. It seems that the phenomenon Kant discovered in the form of the antinomies of pure reason is closely related to the expressive boundaries of the language of science, and in a linguistic form it has a universal validity.

Acknowledgements

The paper was written in the framework of the *Jan Evangelista Purkyně Fellowship* at the Institute of Philosophy of the Academy of Sciences of the Czech Republic.

References

Carnot, S. 1824/1986. *Réflexions sur la Puissance Motrice du Feu*. Paris: Chez Bachelier Libraire. *Reflexions on the Motive Power of Fire*. Manchester: Manchester University Press. 1986.

Cottingham, J. ed. 1992. *The Cambridge Companion to Descartes*. New York: Cambridge University Press.

Descartes, R. 1644/1983. *Principia Philosophiae*. Ámsterdam: Ludovicum Elzevirium. English version by V.R.Miller and R.P. Miller. Dordrecht: Reidel.

Drake, S. 1978. *Galileo at Work: His Scientific Biography*. Chicago: The University of Chicago Press.

Euler, L. 1736. *Mechanica sive motus scientia analytice expósita*. Petropoli: Academia Scientiarum.

Euler, L. 1750. "Découvert d'un principe de Mécanique," in *Mémoires de l'academie des sciences de Berlin*, 6, 185–217. Reprinted in 1957: *Leonhardi Euleri Opera Omnia*, edited by J.O. Fleckenteins, Series Secunda, 5, pp. 81–108. Zurich: Orell Fussli.

Faraday, M. 1831/1955. *Experimental Researches in Electricity*, in Hutchings, M. ed. *Great Books of the Western World*. London: Encyclopedia Britannica.

Fourier, J. 1822. *Théorie analytique de la chaleur*, Paris: Firmin Didot. English version: *The Analytical Theory of Heat*. New York: Dover, 1955.

Gabbey, A. 1980. "Force and Inertia in the Seventeenth Century: Descartes and Newton," in S. Gaukroger ed. *Descartes, Philosophy, Mathematics and Physics*, (pp. 230–320). Sussex: Harvester Press.

Galilei, G. 1623/1957. "*The Assayer*," in S. Drake ed. *Discoveries and Opinions of Galileo*, (pp. 229–280). N. York: Doubleday.

Galilei, G. 1638/1914. *Dialogues Concerning Two New Sciences*. Transl. by H. Crew and A. de Salvio. New York: Macmillan.

Gaukroger, S. ed. 1980. *Descartes, Philosophy, Mathematics and Physics*. Sussex: Harvester Press.

Kant, I. 1781/1990. *Kritik der reinen Vernunft*. Hamburg: Felix Meiner.

Kvasz, L. 2002. "Galilean Physics in Light of Husserlian Phenomenology," in *Philosophia Naturalis* 39, 209–233.

Kvasz, L. 2003. "The Mathematisation of Nature and Cartesian Physics," in *Philosophia Naturalis*, 40, 157–182.

Kvasz, L. 2008. *Patterns of Change, Linguistic Innovations in the Development of Classical Mathematics*. Basel: Birkhäuser Verlag.

Laplace, P.S. 1816. "Sur la vitesse du son dans l'air et dans l'eau," in *Annales de chimie* 3, 238–241.

Mach, E. 1893/1902. *The Science of Mechanics*. Chicago: The Open Court.

Maxwell, J.C. 1861. "On the Physical Lines of Force," in *Philosophical Magazine* 21, 161–175.

Maxwell, J.C. 1873/1954. *A Treatise on Electricity and Magnetism*. New York: Dover.

McMullin, E. ed. 1967. *Galileo, Man of Science*. New York: Basic Books.

Navier, H. 1823. "Mémoire sur le lois du mouvement des fluids," in *Mémoires de l'Académie Royale des Sciences de Paris*, 6, 389–416.

Newton, I. 1670/1988. *Über die Gravitation.* Frankfurt: Vittorio Klostermann.

Newton, I. 1687/1999. *The Principia,* A New Translation by I. Bernard Cohen and Anne Whitman, Proceeded by *A Guide to Newton's Principia.* Berkeley: University of California Press.

Shea, W.R. 1978. "Descartes as Critic of Galileo," in Butts, R. and Pitt, J. eds. *New Perspectives on Galileo,* (pp. 139–160). Dordrecht: Reidel.

Shea, W.R. 1991. *The Magic of Numbers and Motion, The Scientific Career of René Descartes.* Canton, MA: Science History Publications.

Westfall, R.S. 1971. *Force in Newton's Physics.* London: Macdonald.

Westfall, R.S. 1980. *Never at Rest. A Biography of Isac Newton.* Cambridge: Cambridge University Press.

Ethical Limits of Science, Especially Economics

Gereon Wolters

Abstract

I give (I) conceptual clarifications relevant for our argumentation: facts versus norms/values, ethical pluralism versus ethical relativism, moral norms versus juridical norms. It is shown that ethical norms are justified using the principle of universalization: ethical arguments may use only principles to which supposedly everybody could give assent. I then (II) deal with ethical limits to the freedom of science imposed from outside, i.e. legislation (e.g. restrictions on experiments on animals or humans), or from (III) imposed from inside, i.e. science itself (e.g. research moratoria, measures to prevent corruption). I then (IV) turn to economics, showing that the leading neoclassical economical theory is among the causes for the enduring financial and economic crisis. I defend three theses: (1) Neoclassical economics has unethically sold itself as safely explaining and predicting as physics. (2) The models of neoclassical economy are based on value-laden ideological beliefs about free markets and economical agents that are sold as value-free science. (3) Neoclassical experimentation that involves whole countries (like in the completely failed "Chile experiment") and societies is immoral.

Keywords

Ethical limits – science – economics – neoclassical – value-laden

I Conceptual Introduction

In times of Latin as *lingua franca* of science and philosophy our topic today would have been *limites scientiae ethici*. Those of you who have had the good luck to have learned Latin at school might immediately ask: is the genitive *scientiae* a *genetivus subjectivus* or *objectivus*? Or, put differently, does the title ask for ethical limits that are *internal* to the process of science, i.e. ethical limits *of* science, or, are we inquiring, whether ethical limits should be imposed on science *from outside*, i.e. ethical limits *to* science. The short answer is: both. We will see that, on the one hand, the process itself of *conducting science* often raises ethical questions, and that the *application of results* of science, on the other, may pose ethical problems. The typical addressee of the *internal* ethical problems is the scientist him- or herself, while the typical addressee of the

 | DOI 10.1163/9789004325401_006

consequential problems is the society or the lawgiving bodies, respectively. On the basis of this distinction the title of the paper would better read as: "Ethical Limits of and to Science".

Ethics seems to be a topic philosophers, and sometimes also theologians, deal with in a professional way. We might ask, why doesn't one leave to the scientists and doctors themselves the reasoned answer to ethical questions that arise in their respective disciplines? The answer that scientists and physicians often find difficult to accept is that scientific or medical competence is categorically different from ethical competence. Scientific competence relates to the *facts of the world* and delivers *descriptive* results, while ethical competence relates to *norms* and *values* and delivers *evaluative* and *normative* results. In short, science tells us *what there is*, while ethics tells us, *what we should do*, or *which things we should value*. This doesn't exclude that a scientist or doctor may give valuable ethical guidelines. But in doing this they do not make use of their *scientific* or *therapeutic* but rather of their *philosophical* competence. Such competence, however, often is badly missing. The degree of confidence of scientists and doctors in their ethical arguments is often negatively proportional to their quality. This we find, of course, also in philosophy and elsewhere, and not only when it comes to the ethics of science.

In standard philosophical parlance there is an important difference between "ethics" and "morals". "Morals" relate to actually existing rules or norms of conduct of persons or groups. It does not matter whether those rules are "good" or "bad". Thus, one speaks, for example, of the morals of the Mafia, or of the investment banking elite of bank *X*, and at the same time of the morals of the Catholic Church, or rural Lutheran communities in Northern Finland. What these examples have in common and what distinguishes them from "ethics" is their lack of *universal* justification. Sure, the moral rules of the Catholic Church are intended to further the common good, different from those of the Mafia or the banks. But their justification has to finally rely on the existence of God and on the authoritative interpretation of His word by the Church, both of which cannot claim universal assent.

It is the philosophical sub-discipline *ethics* that attempts the justification of moral norms in a universalized form. "Universalization" means: taking recourse to principles and arguments to which supposedly everybody could give assent, provided that one lives with the intention to morally respect other people. Kant has called this intention "the good will" (*der gute Wille*). There is a large variety of attempts to systematize ethics: Kantian ethics is based on the "categorical imperative". One of its formulations is: "act only according to that maxim whereby you can at the same time will that it should become a universal law without contradiction" (Kant 1785/1993, p. 30). So called consequential ethics concentrate on the overall consequences of our actions and are based on some principle of utilitarianism, e.g. Jeremy Bentham's classical definition: "By the principle of utility is meant that principle which approves or disapproves of every action whatsoever according to the tendency it appears to have to augment or diminish the happiness of the party whose interest is in question: or, what is the same thing in other words to promote or to oppose that happiness" (Bentham 1789/2007, p. 1).

These first two really universalizing approaches to ethics have been refined and reformulated in the course of the more than 200 years of their existence. They are at the same time the most explicit examples of ethical universalization. Although others, less explicit ones, exist, like the recourse to Aristotelian virtue ethics, and all sorts of mixed systems, we could, nonetheless, say: ethical norms are universalistically justified moral norms.

As to ethical norms there is a striking similarity to the descriptive realm, which is characteristic of science. As is universally accepted these days, all scientific statements are *as a matter of principle hypothetical*, even the optimally justified and reliable ones. There is, on principle, no absolute or infallible knowledge in the realm of the factual, even if we are convinced that many scientific statements, laws of nature etc. hold firmly without any prospect of ever changing. What applies to the descriptive realm applies also to the normative. Ethical justification is based on principles, which we cannot "prove" in a definite sense – as little as we can "prove" the laws of nature. This is already clear from the fact that there exist various such principles, whose application may lead to diverging moral norms. Apart from that, the application of ethical principles and moral norms is *not an algorithmic procedure* that leads to the same results with everybody. Rather, it rests on *judgment*, and judging has to take into account both the principles *and* the circumstances of their application. The result is what one might call *moral pluralism*. Moral pluralism is the form of morals in secular democratic states, where no institution can claim to be in the possession of absolute truths, be it scientific truths, be it moral truths.

Moral pluralism does not, however, mean moral relativism, because – despite their differences – all moral principles have one thing in common: their *ratio essendi* is the insight that other beings have moral rights towards us. Every universal ethical conception can be regarded as an attempt at developing the norms, which are included in the moral respect that we owe to other beings. This common ground in my view unites different ethical approaches more than their differences separate them.

Another distinction is of great importance in our context: the difference between *ethics* and *law*.[1] Sure, both fields overlap: There are many laws that have an ethical foundation. Think, for example, of those sections of the penal code that forbid murder, fraud, pedophiliac actions and the like. Such sections are the forensic form of moral norms plus the threat of punishment for their violation. But there are moral norms as the imperative not to lie or imperatives in the wide field of partnership that – in general – are not at the same time legal norms. Another difference between moral norms and legal norms relates to *conviction*. Legal norms require simply a certain *behavior*. It is irrelevant, whether one takes a legal norm to be reasonable or nonsense as long as one behaves according to that norm. Take e.g. speed limits. It does not matter whether you deem speed limits as severe restriction of your freedom, as long as you keep within the speed limit. It is hardly imaginable, however, that somebody speaks the truth even to his/her disadvantage, but at the same time reckons the moral norm "you shall no lie!" to be mistaken.

Most important in the context of morals and law is the question, which moral norms should be protected by law. As to be expected, there exist different answers to this question. I very much support the enlightenment conception, which includes that religion is a private affair, and that, accordingly, religion and state should be kept separate from each other.[2] This includes that norms based on religious belief do not have any privilege in the political discussion about legal sanctions of moral norms. Furthermore, I support the liberal conception that the democratic state should interfere with the private concerns of citizens as little as possible and as much as necessary. This leads to the answer that moral norms need legal sanctioning only if they express a *common* good whose implementation is vital for the functioning of society.

II Moral Limits Imposed on Science

After this long conceptual overture we have, finally,properly arrived at our topic. Let us first have a look at moral limits that are imposed on science from outside, i.e. by law. Putting morally justified legal limits to science means,

1 Much that relates to this topic is drawn from Wolters (1991).

2 In Wolters (2013) I have given an analysis of the relationship between religion and enlightenment. Different from France in countries like Germany the Churches hold still remarkable privileges. In Germany they include the restriction of the freedom of research and teaching at state universities (see Section IV of that chapter).

first of all, restricting academic freedom. Probably in all European countries academic freedom is guaranteed either by law or by the constitution. The German Constitution of 1949 (*Grundgesetz*) in its first part on the fundamental rights of the citizens succinctly states: "Art and science, research and teaching are free. The freedom of teaching does not absolve from the allegiance to the constitution."[3] Similarly the Spanish Constitution of 1978 states in article 20.1c of its first section "De los derechos fundamentales y de las libertades públicas": "Se reconocen y protegen los derechos: [...] a la libertad de cátedra."[4]

The status of academic freedom as a fundamental right implies that possible restrictions of this right need convincing justifications. Restrictions of academic freedom exist in every European country. Although I can speak here only about the German case, I am pretty sure that things in Spain are not entirely different. Here a few examples: there are restrictions of research in order to protect animals. To the best of my knowledge there are laws, based on moral considerations, in every European country that restrict animal experimentation. Much stricter laws hold for experiments on humans, which, in addition, are regulated by international declarations. Such declarations elaborate and adjust the first such declaration of Helsinki of 1964 to new circumstances.

Another example of moral restrictions of research relates to genetic cloning of humans. There are people who are so sure of themselves that they would like to genetically multiply. As we know, the technique of genetic cloning works in the animal kingdom. Some of you might remember the first cloned mammal: Dolly, the domestic sheep that was born as a clone of another female sheep. Why not clone humans? Why not have a second and even more editions of Silvio Berlusconi, Lady Gaga or Lionel Messi? Sure, we know from epigenetics and from the importance of culture in human development that such clones would not be as identical with the original as the latter might want. But in any case, some spectacular similarity would come about. To the best of my knowledge there does not yet exist a human clone. The legal situation in Europe is somewhat confusing. In Spain cloning is prohibited by the European Convention on Human Rights and Biomedicine that has been ratified by Spain, while most other European countries have not joined so far.[5]

One can adduce here various moral reasons for prohibiting reproductive cloning of humans. I would like to mention only two: First of all, one would need *experiments* in order to establish procedures. As everybody knows, experiments can go wrong and deliver undesired results. How about a cloned

3 Basic Law of the Federal Republic of Germany (1949, article. 5, Section 3).

4 Constitución española (1978).

5 This results at least from the Wikipedia article "Human Cloning" (seen March 2014).

baby that is born with a severe handicap? Furthermore, as with Dolly, there is the possibility that also a human clone, which is seemingly born healthy, develops ailments over time that are related to his or her being a clone. Describing such cases gives already an answer to our question about restricting academic freedom in this area: Cloning is morally excluded by simple moral principles.

The restrictions I have talked about so far, i.e. restrictions concerning experiments on animals or humans are restrictions that directly relate to the *process* of research. Many more moral problems arise, however, when it comes to the *application* of research. In Germany for many years there has been a vivid public discussion about preimplantation genetic diagnosis (PGD), which was forbidden until recently by the Embryo Protection Law of 1990, while in most other European countries PGD was practiced without legal problems. What is at issue?

PGD is a diagnostic procedure that allows genetic screening of an embryo generated by *in vitro*-fertilization, before it is implanted. It is used in cases in which there exists a high risk that a baby will be born with a severe hereditary disease. The ethical questions that arise in the context of PGD are basically the same as in the case of abortion. Minor issues concern the question of surplus embryos or the valuation of handicapped life.

Note that any restriction in the case of PGD does not relate to scientific research but rather to its application. Similar questions arise in other fields. Take atomic research as it is applied by the atomic industry in order to construct atomic power plants. Again, atomic research itself is "innocuous", its application is not. To mention just one point: atomic waste. Plutonium-239 that is generated in reactors has a half life of 24,000 years. Note that we can identify the first Egyptian cultures some 6,000 years ago. Could they properly communicate with us, as we now should be able to communicate with cultures several times farther ahead of us, and tell them how they have to treat nuclear waste? I doubt this very much. Apart from this there is the moral problem of future generations that will live much closer to us: they might have to pay the price for our way of life.

III Moral Limits Set by Scientists Themselves

In this section I would like to talk about moral limits that are imposed on science not from outside but, rather, set by researchers themselves. This may occur on the individual as well as on the institutional level. Somebody might opt out from a certain type weapon related research, e.g. chemical weapons, because he/she objects to the use of such weapons for moral reasons. Others might leave military research altogether because for them wars in general are morally unjustified. There are possibly more such individual pacifist options than we might hear of in the media.

An interesting historical example is the German *Uranprojekt*, which from 1939 on tried to lay the scientific foundations for building an atomic reactor and a bomb. Particularly physicists Werner Heisenberg and Otto Hahn seem to have had great hesitations to build the bomb for the Nazi government. This results from intercepted conversations of the German scientists in Farm Hall (England), where they had been detained after the war by the British secret service (Hoffmann 1993). Particularly Hahn, who in 1938 had detected nuclear fission in 1938, felt personally co-responsible for the death of more than 90,000 people in Hiroshima, where the first (American) atomic bomb was dropped.

A more recent example of a research moratorium on an *institutional* level are the guidelines that were worked out at a conference on recombinant DNA in Asilomar (California) in 1975. This had to do with the potential danger of creating deadly monsters by genetically modifying existing ones. The guidelines forbade, in fact, certain types of potentially dangerous experiments. The Asilomar conference and its guidelines turned out to be a milestone in the development of interaction between biological science and society. Scientists became more and more aware that they owe responsibility for their work to the society that finances it. Financing leads us to another problem in the context of ethical limits of research.

I would like to mention here only one point that is related to the problem that much research is not financed by the state or public institutions but by private companies. Private companies do not act for philanthropic reasons. They would like to see a quick return for the money they invest. One could say, of course, there is no problem: science delivers objective results. Therefore, it is of no importance who finances research. This is, unfortunately, not so. Objectivity is, in fact, one of the ideals of science. It is an *ideal*, though, that is often realized only in a rather approximate way.[6] There are several epistemic parameters of research projects, where values und judgments of the researchers enter, and with them possibly the interests of the sponsors. The researchers are in many cases certainly not aware of this influence, whose existence has been proved in many cases. In a paper of 1986 Richard A. Davidson has studied 107 controlled clinical trials, in which a traditional drug therapy and a therapy with new drugs were compared (Davidson 1986, pp. 155–158. In the context of funding see also Brown 2008). The 107 studies were classified in two

6 Cf. Introduction and many articles in Machamer, P. and Wolters, G. (2004) and in Carrier, M. et al. (eds.) (2008).

ways. First, whether they favored the new drugs or the traditional drugs, and second, whether the trials were financially supported by a pharmaceutical firm or by public money. The result: "The study has demonstrated a statistically significant association between source of funding (pharmaceutical firm versus general support) and outcome of the published clinical trials" (Davidson 1986, p. 156f). Although this study refrains from establishing a causal connection, everything suggests that the interests of the sponsors, perhaps without the conscious intention of the researchers, somehow diffused into the result of research.

There are more recent examples that seem to be less innocent. I rather think that they point into the direction of corruption and/or ideology. In this context Naomi Oreskes' and Eric Conway's book *Merchants of Doubt* is of utmost importance (Oreskes and Conway 2010). Here is a quote from the website of the book: "The U.S. scientific community has long led the world in research on public health, environmental science, and other issues affecting the quality of life. Our scientists have produced landmark studies on the dangers of DDT, tobacco smoke, acid rain, and global warming. But at the same time, a small yet potent subset of this community leads the world in vehement denial of these dangers. [...] Naomi Oreskes and Erik Conway explain how a loosely–knit group of high-level scientists, with extensive political connections, ran effective campaigns to mislead the public and deny well-established scientific knowledge over four decades. In seven compelling chapters addressing tobacco, acid rain, the ozone hole, global warming, and DDT, Oreskes and Conway roll back the rug on this dark corner of the American scientific community, showing how the ideology of free market fundamentalism, aided by a too-compliant media, has skewed public understanding of some of the most pressing issues of our era."

In my view there is no question that those scientists, whom Oreskes and Conway address, and many others they did not talk about, have severely violated the ethics of scientific research. There are, indeed, moral limits scientists ought to put themselves, in order to preserve both the ideal of scientific objectivity and the well-being of their society and of the whole world.

The expression "free market fundamentalism" in the above quote brings me to the last section of the paper.

IV Ethical Limits of Science – Largely Ignored by Economists

Commencing in 2007, Western countries have been experiencing an enormous economic crisis, Spain being one of those hit hardest. The crisis began as a crisis of financial markets triggered by the U.S. real estate bubble, the bankruptcy of the Lehman Bank, the almost meltdown of the AIG insurance giant, and similar disasters.[7] Quickly, real economy was affected with devastating social consequences.

There exists a "Financial Crisis Inquiry Report" of 662 pages that the "National Commission on the Causes of the Financial and Economic Crisis in the United States"[8] presented to the US government in January 2011. Its "Conclusions" about the causes of the crisis, which has been judged as "avoidable", are as follows: "[…] widespread failure in financial regulation and supervision proved devastating to the stability of the nation's financial markets. […] dramatic failures of corporate governance and risk management at many systematically important financial institutions were a key cause of this crisis. […] a combination of excessive borrowing, risky investments, and lack of transparency put the financial system on a collision course with crisis. […] We conclude the government was ill prepared for the crisis, and its inconsistent response added to the uncertainty and panic in the financial markets. […] there was a systematic breakdown in accountability and ethics. […] collapsing mortgage-lending standards and mortgage securitization pipeline lit and spread the flame of contagion and crisis. […] over-the-counter derivatives contributed significantly to this crisis. […] the failures of credit rating agencies were essential cogs in the wheel of financial destruction."

The "Conclusions" conclude: "There is still much to learn, much to investigate, and much to fix. This is our collective responsibility. It falls to us to make different choices if we want different results."[9]

What is fascinating about this report is that nobody in the US government seems to be interested in a *possible scientific background* of the glamorous

7 A fascinating analysis gives Stiglitz (2010). Stiglitz is one of the winners of the Nobel Prize in economics in 2001, and at the same time one of the most thorough critics of the ruling neoclassical paradigm (see below).

8 The "Conclusions" as well as the entire report are available at: fcic.law.stanford.edu/report/conclusions.

9 Although the word "fraud" occurs in the report "no fewer than 157 times", interestingly, not one high level executive has been prosecuted so far. Cf. Rakoff (2014). The author, a United States District Judge for the Southern District of New York – Wall Street is situated there – gives fascinating answers to the title question. Among them is a juridical parallel for the political-economic "too big to fail": too big to jail. The situation in Europe is certainly not very different.

failure of economic policy, i.e. a possible background in mainstream economic theory, otherwise called “neoclassical economics”.[10]

There are, however, highly respected economists, who see things differently. Joseph E. Stiglitz writes: “As we peel back the layers of >what went wrong<, we cannot escape looking at the economics profession. Of course, not all economists joined in the jubilation of free market economics; not all were disciples of Milton Friedman. A surprisingly large fraction, though, leaned in that direction. Not only was their advice flawed; they failed in their basis tasks of prediction and forecasting. [...] It was not an accident that those who advocated the rules that led to the calamity were so blinded by their faith in free markets that they couldn't see the problems it was creating. *Economics had moved* – more than economists would like to think – *from being a scientific discipline into becoming free market capitalism's biggest cheerleader* [my emphasis]” (Stiglitz 2010, p. 238).

Such a move from scientific theory to political cheerleader, unsurprisingly, has ethical implications as to how responsibly conduct economical theory. I would like to argue for three interconnected theses:

Thesis 1: The attempt of mainstream neoclassical economics to model itself exclusively after physics is misguided and has failed. It is, therefore, unethical to sell economical theory to the public as safely explaining and predicting.

Thesis 2: The failure of neoclassical economics rests largely on confounding abstract mathematical models with reality. Those models are based on ideological beliefs about free markets and on the conception of the economic agents as rational maximizers of their own utility (*Homo oeconomicus*). The

10 In this scientific context I avoid the word “neoliberalism” that has become a poorly defined “academic catchphrase” and a political combat term. On the history of “neoliberalism”, cf. Boas and Gans-Morse (2009). There is a vast and polemical anti-neoliberalism literature in all major languages. Particularly interesting is in my view Mirowski and Plehwe (2009). Various contributions to the book show remarkable differences in “liberal” approaches to economic theory, before the “Chicago School” (Milton Friedman, and others) succeeded to establish a sort of “imperialistic” paradigm that more and more has determined mainstream economics in the rest of the world (especially Ch. 4, pp. 139–178). Very precious insights in the thinking and acting of leading American economists, including deans of prestigious Business Schools, who often got the bulk of their income from well paid consulting and even regulating jobs, gives the documentary film Ferguson, C. 2010. *Inside Job*, SONY Pictures. I would like to highly recommend this movie.

connected ethical failure is the lack of reflecting the value implications of one's own approach that is sold as value-free science.

Thesis 3: Neoclassical experimentation that involves whole countries and societies is immoral.

Ad thesis 1: Neoclassical economics is something like the common background of almost everybody who at present works as an economist with Western governments, banks, firms etc. What physics, chemistry and materials science is for engineers, neoclassical economics is for business people, regulators and business related governments, and has led economists to regard their discipline as a sort of social physics. Correspondingly, critics of this approach often speak of "physics envy". Physics envy, generally, stands for the desire of mainstream social science, including above all economics, to become a science like physics: fully mathematized and reliable like mechanics, electromagnetism or quantum theory. This desire might be fueled in Anglo-Saxon countries by the linguistic peculiarity that "science" normally only relates to what elsewhere is usually called "natural science". This seems to have led to a sort of inferiority complex among economists and other social "scientists", which they tried to overcome above all by mathematizing their disciplines. Economics has been certainly most successful in approaching this ideal. But not only mathematics is required in order to become a true "science". What one needs, furthermore, according to aspiring *real* social scientists, is the hypothetical deductive method. The hypothetical deductive method is typical of natural science, but certainly not the only respectable method in town. It includes first of all constructing theoretical models, then giving testable hypotheses derived from such models, and, finally, empirical tests of the hypotheses against the reality of the social world.

I am certainly not criticizing the attempt of economics to model itself after physics, as long as two important restrictions are kept on record. *Firstly*, this methodological move is not necessary, although the social sciences, particularly in Anglophone countries, have almost completely adopted such modeling as the only respectable way of conducting functioning social science. Modeling economics after physics neglects, among other things, the fact that social sciences have successfully worked with models that were never tested empirically (Clark and Primo 2012). Furthermore, social sciences have even effectively worked without models, e.g. with analytical narratives. *Secondly*, there is, moreover, a remarkable discrepancy between pretensions and reality. It has always been the aim of physics, and is, in fact, included in the hypothetical deductive model, to deliver "precise explanations and successful predictions" (Gillies 2012, pp. 351–362, especially p. 355). To his great surprise, philosopher Donald Gillies, when checking the textbooks of four famous Nobel Prize winning neoclassical economists (Samuelson, Arrow, Debreu, Prescott), found that they achieved neither. This led Gillies to the "conjecture that the use of mathematics in mainstream (or neoclassical) economics has not produced any precise explanations or successful predictions. This [...] is the main difference between neoclassical economics and physics, where both precise explanations and successful predictions have often been obtained by the use of mathematics" (Gillies 2012, p. 362).

This seems to me a rather sobering result for pretensions to model economics after physics. It points to an overassessment of the mathematical structure and predictive power of economics. It might well be that the initial and boundary conditions of economical models and the changes over time of the latter are so complex that they cannot be effectively modeled to the desired degree, at least for the time being. It might well be that economics is more like meteorology, or "like medicine" as economist Dennis Snower recently put it in an interview (2014).[11] – Similarly, Jean-Claude Trichet, president of the *European Central Bank* (ECB) from 2003 to 2011, who had to deal in the front rank with the disastrous consequences of the crisis, summarizes: "In the crisis, finally, the available economic and financial models, above all the dynamical stochastic models of general equilibrium, were largely not working."[12]

Failure in explaining and predicting is not special to economics. Rather, failure is part of the normal course of science. Special to economics is, however, a possible ethical dimension of that failure. It seems obvious that many economists failed to honestly convey to their students, to politicians, and to the general public the limited explanatory and predictive scope of their models, compared to those of physics. They should have known better.

There are remarkable exceptions, though, e.g. the Israeli economist Ariel Rubinstein. In the introduction to lecture notes he tells his students: "I do not view economic models as an attempt to describe the world or to provide tools for predicting the future. I object to looking for an ultimate truth in economic

11 Snower is the director of the "Institut für Weltwirtschaft" at Kiel, Germany and gave the interview on the occasion of the centenary celebration of the institute.

12 "Enfin, dans la crise, *les modèles économiques et financiers disponibles* – en particulier les modèles dynamiques stochstiques d'équilibre général – se sont avérés largement inopérants," Trichet (2013, p. 251).

theory, and I do not expect it to be the foundation for any policy recommendation. Nothing is >holy< in economic theory and everything is the creation of people like yourself. […] The word >model< sounds more scientific than >fable< or >fairy tale<, but I don't see much difference between them. […]. The fable is an imaginary situation that is somewhere between fantasy and reality. Any fable can be dismissed as being unrealistic or simplistic, but this is also the fable's advantage. […] a good model in economic theory, like a good fable, identifies a number of themes and elucidates them" (Rubinstein 2012, pp. ix–x).

Whatever the exact methodological value of the comparison fairy tales/models may be, it seems to me important that students of economics are unequivocally informed that they are not studying physics.

Ad thesis (2) To this failure I would like to add another one: confounding abstract models with social reality, and at the same time not reflecting the value implications at the basis of those models. First of all, neoclassical economic models do not deal with real human beings. Their agent is the *Homo oeconomicus*, a creation of economists.[13] He is understood as a rational agent that acts exclusively in order to maximize his own utility. In this context human beings are interesting only as "consumers and firms. We assume that consumers seek to maximize utility and that firms seek to maximize economic profit, which is the difference between total revenue and total cost" (Rittenberg and Tregarthen 2010, Ch. 6, p. 2).

The most important instruments of *homo oeconomicus* in maximizing his utility are *rational choice theory* and *game theory*. Secondly, the most important boundary condition of the neoclassical approach is perhaps the so called *Efficient Market Hypothesis* (EMH), which states that agents on (financial) markets act completely rational on the basis of equal information. This is said to lead to a stable market equilibrium. Other boundary conditions include non-intervention from government and non regulation of markets with the exception of the regulation of money supply by the central banks. These boundary conditions are sometimes polemically called "market fundamentalism". This is the world, where *Homo oeconomicus* has come to life.

13 On the concept of *Homo oeconomicus* cf. the encompassing analysis of Kirchgässner (2008). Kirchgässner emphasizes the distinction between the *methodological* value of the *Homo oeconomicus* model, and the *metaphysical* assumption that people, in fact, act exclusively out of self-interest.

To be sure, using the model of *Homo oeconomicus* in the social sciences can create great insights, as long as one is clear about the fact that it is a methodological instrument of research and does not describe social reality in an encompassing and reliable way. Exactly this distinction between model and reality has come out of sight among many economists,[14] and *Homo oeconomicus* seems to have become reality in stock exchanges, banks, investment firms, and so on. But there is no doubt: "*Homo oeconomicus* is a sociopath", as the Cornell jurist Lynn A. Stout has succinctly put it (Stout 2008, p. 159). And Joseph Stiglitz ironically observes: "One interesting aspect of economics is that the model [of *Homo oeconomicus*] provides a better description of economists than it does of others, and the longer students study economics, the more like the model they become" (Stiglitz 2010, p. 249).[15]

Bringing the *Homo oeconomicus* alive means selling abstract models as social reality. In other words, many mainstream neoclassical economists do not ask any more how humans are *really* wired. Rather, they claim that humans are wired just as their models assume, i.e. humans are beings that are constantly and permanently maximizing their own utility.

This is not without importance for the ethics of research. Selling the *Homo ooeconomicus* in his market environment as real is not only a scientific but also an ethical failure. Allegedly value-free science is imbued with political values, basically the political market fundamentalism, preached with missionary zeal of American republicans and libertarians, particularly of the Reagan era. It soon made its way to Margaret Thatcher's Great Britain and, slightly attenuated to other countries.[16] The rather enlightened American billionaire George Soros has put it this way: "Fundamentalists believe that markets tend towards equilibrium and the common interest is best served by allowing participants to pursue their self-interest. It is an obvious misconception, because it was the intervention of the authorities that prevented financial markets from breaking down, not the markets themselves. Nevertheless, market fundamentalism emerged as the dominant ideology in the 1980s, when financial markets started to become globalized and the US started to run a current account deficit" (Soros 2008).

---

14 Think of Jean-Claude Trichet's quote above. Trichet devotes to "L'hypothèse des marchés efficient" a particularly trenchant analysis. See Trichet (2013, pp. 256–258).

15 Stiglitz' remark is based on a study about attitudes of students of economy.

16 I am not criticizing market economy as such, but only its fundamentalist variant.

This fundamentalism has been, nonetheless, sold as science by mainstream neoclassical economy. This is not only a scientific shortcoming. Responsible scientists are also morally obliged to reflect the value presuppositions of their theories, render them explicit, and to change them, when they empirically fail or are in conflict with moral principles or principles of justice.

Ad thesis (3) My last point relates to *experiments* in economics. Wenceslao Gonzalez has written on ontological, methodological and epistemological aspects of various types of such experiments, focusing on the position of the German Nobel prize winner Reinhard Selten, one of the dissidents from neoclassical orthodoxy, who rejects the strong rationality assumption of that school (Gonzalez 2007). I would like to deal here with *ethical* problems that may result from the experimental application of neoclassical economy.

As an example I take the attempt of the "Chicago Boys" to turn upside down the economy of Chile.[17] The "Chicago Boys" were a strongly connected group, whose core consisted of some twenty Chilean economists, most of whom had studied economics with Milton Friedman and others at the University of Chicago. The "Boys" travelled from Chile to Chicago from the mid fifties onward thanks to a cooperation program of Chicago University with the *Universidad Católica* in Santiago de Chile. This is not without irony because the Catholic Social Teaching, based on the principle of justice, is hardly reconcilable with the economic and social recipes issued by neoclassical economics.

Under Augusto Pinochet's dictatorship the time had come for the Chilean "Chicago Boys" and their Chicago mentors to replace Chile's protectionist economy by American style neoclassical economy. In 1975, Pinochet appointed four "Chicago Boys" as minister of economics, minister of finance, president of the Central Bank and president of the Office of Economic Planning, respectively. The "Great Chile experiment" could start. It included privatization of the public economic sector and of the health and pension systems. Strikes were forbidden, taxes reduced, and so on. To cut a complex story short: the "Great Chile experiment" that put neoclassical economy to test, was, finally, a failure, and resulted in a collapse of the Chilean financial market in 1982. Sixteen of 50 private financial institutions went bankrupt with the usual consequences for real economy. To suffer were above all the poor.

17 There is an excellent article "Chicago Boys" in the German Wikipedia (seen on March 2014).

We have seen that there are strict regulations in medicine for experiments on humans. The same does, unfortunately, not hold for economics. Ideology laden economists can do extreme social harm to people, when they are allowed to apply their theories in grand experiments as happened in the case of Chile. Protected by a dictator of a third world country they could just go ahead as their American ideological agenda suggested. If there had been democratic control, such experiments would hardly have been possible.

In this context we might reflect how much, or better, how little democratic control of the financial markets is still left to sovereign states in our days.

References

Basic Law of the Federal Republic of Germany (1949), available at: http://en.wikisource.org/wiki/Basic_Law_for_the_Federal_Republic_of_Germany, (accessed on 5.4.2016).

Boas, T.C. and Gans-Morse, J. 2009. "Neoliberalism: From New Liberal Philosophy to Anti-Liberal Slogan," in *Studies in Comparative International Development*, 44, 137–161.

Bentham, J. 2007. *An Introduction to the Principles of Morals and Legislation* [1789]. Mineola, NY: Dover Publications.

Brown J.R. 2008. "The Community of Science," in Carrier, M., Howard, D. and Kourany, J. eds. *The Challenge of the Social and the Pressure of the Practice: Science and Values Revisited*, (pp. 189–216). Pittsburgh: University of Pittsburgh Press.

Carrier, M., Howard, D., and Kourany, J. eds. 2008. *The Challenge of the Social and the Pressure of the Practice: Science and Values Revisited*. Pittsburgh: University of Pittsburgh Press.

Clark, K.A. and Primo, D.M. 2012. "Overcoming 'Physics Envy': How Scientific are the Social Sciences?" in *New York Times*, March 30, 2012, available at: http://www.nytimes.com/2012/04/01/opinion/sunday/the-social-sciences-physics-envy.html, (accessed on 5.4.2016).

Constitución española de 1978, available at: http://es.wikisource.org/wiki/Constituci%C3%B3n_espa%C3%B1ola_de_1978, (accessed on 5.4.2016).

Davidson, R.A. 1986. "Source of Funding and Outcome of Clinical Trials," in *Journal of General Internal Medicine*, 12 (3), 155–158.

Ferguson, C. 2010. *Inside Job*, SONY Pictures (Film).

Gillies, D. 2012. "The Use of Mathematics in Physics and Economics: A Comparison," in Dieks D. *et al.* eds. *Probabilities Laws, and Structures*, (pp. 351–362). Dordrecht: Springer.

Gonzalez, W.J. 2007. "The Role of Experiments in the Social Sciences: The Case of Economics," in Kuipers, T. ed. *Handbook of the Philosophy of Science – Focal Issues*, (pp. 275–301). Amsterdam: Elsevier.

Hoffmann, D. ed. 1993. *Operation Epsilon. Die Farm-Hall-Protokolle oder die Angst der Alliierten vor der deutschen Atombombe*. Berlin: Rowohlt.

Kant, I. 1785. *Grounding for the Metaphysics of Morals*. English translation by J.W. Ellington, Hackett, Cambridge Mass., 3rd edition 1993.

Kirchgässner, G. 2008. *Homo Oeconomicus: The Economic Model of Behaviour and Its Applications in Economic and Other Social Sciences*. New York: Springer.

Machamer, P. and Wolters, G. eds. 2004. *Science, Values, and Objectivity*. Pittsburgh and Konstanz: University of Pittsburgh Press and Universitätsverlag.

Mirowski, P. and Plehwe, D. eds. 2009. *The Road from Mount Pelerin: The Making of the Neoliberal Thought Collective*. Cambridge, MA: Harvard University Press.

National Commission on the Causes of the Financial and Economic Crisis in the United States, *Financial Crisis Inquiry Report*, Washington, 2011, available at: fcic.law.stanford.edu/report/conclusions (accessed on 5.4.2016).

Oreskes N. and Conway, E. 2010. *Merchants of Doubt: How a Handful of Scientists Obscured the Truth on Issues from Tobacco Smoke to Global Warming*. New York, Berlin and London: Bloomsbury

Rakoff, J.S. 2014. "The Financial Crisis: Why have no High-Level Executives Been Prosecuted?" in *The New York Review of Books*, 61 (1), (available at the *NYR* website).

Rittenberg, L.and Tregarthen, T., *Principles of Microeconomics*, Webversion 2010, available at: http://www.web-books.com/eLibrary/NC/Bo/B63/TOC.html, (accessed on 5.4.2016).

Rubinstein, A. 2012. *Lecture Notes in Microeconomic Theory: The Economic Agent*. Princeton and London: Princeton University Press, (2nd edition).

Snower, D. 2014. Wir Ökonomen sind Ärzte", *Frankfurter Allgemeine Sonntagszeitung*, January 19.

Soros, G. 2008. "The Worst Market Crisis in 60 Years," in *Financial Times*, January 22. Available at: http://www.ft.com/intl/cms/s/0/24f73610-c91e-11dc-9807-000077b07658.html#axzz2N3WxuGO4, (accessed on 5.4.2016).

Stiglitz, J.E. 2010. *Freefall. Free Markets and the Sinking of the Global Economy*. New York: W.W. Norton.

Stout, L.A. 2008. "Taking Conscience Seriously," in *Moral Markets: The Critical Role of Values in the Economy*, (pp. 157–172). Princeton: Princeton University Press.

Trichet, J.-C. 2013. "Aujourd'hui: la théorie face à la crise," in Buser, P., Debru, C. and Meyer, P. eds. *Les Lumières: hier aujourd'hui, demain. Science et societè – Die Aufklärung: gestern, heute, morgen. Wissenschaft und Gesellschaft. Colloque interacadémique franco-allemand 7 et 8 février 2013, à l'occasion du cinquantième anniversaire du Traité de l'Élysée*, (pp. 249–265). Paris: Hermann Éditeurs.

Wolters, G. 1991. "Einschränkungen der Forschungsfreiheit aus ethischen Gründen?" in Holzhey, H., Jauch, U.P., and Würgler, H. eds. *Forschungsfreiheit. Ein ethisches und politisches Problem der modernen Wissenschaft*, (pp. 199–214). Vdf, Zürich.

Wolters, G. 2013. "Aufklärung und Religion – damals und heute," in Buser P., Debru C., and Meyer P. eds. *Les Lumieres: hier, aujourd'hui, demain – Science e societè. Colloque interacadémique franco-allemand*, (pp. 219–238). Paris: Hermann Éditeurs.

PART 3

Epistemological Limits to Science

Predicting and Knowability: The Problem of Future Knowledge

Nicholas Rescher

Abstract

Predicting the future knowledge poses drastic problems. In this regard, there is a difference between details and generality. In addition, there are several epistemological problems: (a) is future knowledge diminishing?; and (b) is future knowledge perfectible? In this context, we can think of erotetic completeness, predictive completeness, pragmatic completeness, and temporal finality.

The problem of knowability concerning future science can be seen in its relation to reality. This involves also taking into account epistemic vistas, which eventually leads to the nexus between science and reality as well as to the ramifications of impredictability. In this regard, future knowledge will in substantial measure be inscrutable for us today. Consequently, the paper considers knowability and examines its wider implications for the condition of human knowledge.

Keywords

Predicting – knowability – future knowledge – perfectability (of knowledge) – scientific realism

1 Predicting Future Knowledge

Philosophers since Aristotle have stressed that knowledge about the future poses drastic problems.[1] And even the issue of knowing the future of knowledge itself is particularly challenging. No-one can possibly predict the details of tomorrow's discoveries today. To be sure, there is no inherent problem about predicting *that* certain discoveries will be made. But their nature is bound to be unfathomable. After all, if we knew the details and could solve tomorrow's problems today, then they simply would not be tomorrow's problems.

1 That contingent future developments are by nature cognitively intractable, even for God, was a favored theme among the medieval scholastics. On this issue see Adams (1987, ch. 27).

 | DOI 10.1163/9789004325401_007

We may assume – or suppose – that homo sapiens will continue to exist and to do so in a form that will enable him to pursue the prospect of inquiry into the nature of things. With this supposed, we do actually know some important facts about the body of knowledge that will be available to the knowers of the future. One of these relates to retrospective knowledge: knowledge of particular facts regarding the past. That Caesar crossed the Rubicon, that Napoleon lost at Waterloo, that Hitler led Germany to invade Poland in 1939, that there were 48 states in the continental US in the year 2000 – these and their like are parts of our currently available body of knowledge that will continue in place in the future. And another of these preserved kinds of knowledge relates to various trans-temporal facts: the speed of sound, the specific gravity of lead, the molecular structure of water, the evolutionary history of man. Such facts will continue securely in place – at least in a rough and approximate formulation.

However, that aspect of the future which is most evidently unknowable is the future of invention, of discovery, of innovation – and particularly in the case of science itself. As Immanuel Kant insisted long ago that every new discovery opens the way to others, every question that is answered gives rise to yet further questions to be investigated.[2] And the fruits of future science are not yet ripe for present picking

The landscape of natural science is ever-changing: innovation is the very name of the game. Not only do the theses and themes of science change but so do the very questions. Scientific inquiry is a creative process of theoretical and conceptual innovation; it is not a matter of pinpointing the most attractive alternative within the presently specifiable range, but one of enhancing and enlarging the range of envisageable alternatives. Such issues pose genuinely open-ended questions of original research: they do not call for the resolution of problems within a preexisting framework but for a rebuilding and enhancement of the framework itself. Most of the questions with which present-day science grapples could not even have been raised in the state-of-the-art that prevailed a generation ago. It is in principle infeasible for us to tell now but only how future science will answer present questions but even what questions will figure on the question agenda of the future, let alone what answers they will engender. In this regard, as in others, it lies in the inevitable realities of our condition that the details of our ignorance are – for us at least – hidden away in an impenetrable fog of obscurity.

And so, the contrast between present knowledge and future knowledge is clearly one that we cannot characterize in detail. It would be utterly unreasonable to expect prognostications of the particular *content* of scientific

2 On this theme see Rescher (2000).

discoveries. It may be possible in some cases to speculate *that* science will solve a certain problem in the future, but *how* it will do so lies beyond the ken of those who antedate the discovery itself. If we could *predict* discoveries in detail in advance, then we could *make* them in advance.[3] In matters of scientific importance, then, we must be prepared for surprises. Commenting shortly after the publication of Frederick Soddy's speculations about atomic bombs in his 1950 book *Science and Life*, Robert A. Millikan, a Nobel laureate in physics, wrote that "the new evidence born of further scientific study is to the effect that it is highly improbable that there is any appreciable amount of available subatomic energy to tap."[4] In science forecasting, the record of even the most qualified practitioners is poor. For people may well not even be able to conceive the explanatory mechanisms of which future science will make routine use.

In inquiry as in other areas of human affairs, substantial upheavals can come about in a manner that is sudden, unanticipated, and sometimes unwelcome. Major scientific breakthroughs often result from research projects that have very different ends in view. Louis Pasteur's discovery of the protective efficacy of inoculation with weakened disease strains affords a striking example. While studying chicken cholera, Pasteur accidentally inoculated a group of chickens with a weak culture. The chickens became ill, but, instead of dying, recovered. Pasteur later reinoculated these chickens with fresh culture – one strong enough to kill an ordinary chicken. To Pasteur's surprise, the chickens remained healthy. Pasteur then shifted his attention to this interesting phenomenon, and a productive new line of investigation opened up. In empirical inquiry, we generally cannot tell in advance what further questions will be engendered by our endeavors to answer those on hand. New scientific questions arise from answers we give to previous ones, and thus the issues of future science simply lie beyond our present horizons.

It is a key fact of life that ongoing progress in scientific inquiry is a process of *conceptual* innovation that always places certain developments outside the cognitive horizons of earlier workers because the very concepts operative in their characterization become available only in the course of scientific discovery itself. (Short of learning our science from the ground up, Aristotle could have made nothing of modern genetics, nor Newton of quantum physics.) The

---

3 As one commentator has wisely written: "But prediction in the field of pure science is another matter. The scientist sets forth over an uncharted sea and the scribe, left behind on the dock is asked what he may find at the other side of the waters. If the scribe knew, the scientist would not have to make his voyage," Anonymous (1920, p. 321).

4 Quoted in *Daedalus*, v. 107 (1978), p. 24.

major discoveries of later stages are ones which the workers of a substantially earlier period (however clever) not only have failed to make but which they could not even have *understood*, because the requisite concepts were simply not available to them. Thus, it is effectively impossible to predict not only the answers but even the questions that lie on the agenda of future science. Detailed prediction is beyond the reach of reasonable aspiration in those domains where innovation is preeminently conceptual.

2 Detail vs. Generality

Forecasts of scientific developments conform to the vexatious general principle that, other things being equal, *the more informative a forecast it, the less secure it is, and conversely, the less informative, the more secure it is*. It is a fundamental law of epistemology that increased confidence in the correctness of our estimates can always be secured at the price of decreased accuracy. For in general an inverse relationship obtains between the definiteness or precision of our information and its substantiation: detail and security stand in a competing relationship. We estimate the height of the tree at *around* 25 feet. We are *quite sure* that the tree is 25 ± 5 feet high. We are *virtually certain* that its height is 25 ± 10 feet. But we can be *completely and absolutely sure* that its height is between 1 inch and 100 yards. Of this we are "completely sure" in the sense that we are "absolutely certain," "certain beyond the shadow of a doubt," "as certain as we can be of anything in the world," "so sure that we would be willing to stake your life on it," and the like. For any sort of estimate whatsoever there is always a characteristic trade-off relationship between the evidential *security* of the estimate, on the one hand (as determinable on the basis of its probability or degree of acceptability), and on the other hand its contentual *detail* (definiteness, exactness, precision, etc.).

An ironic but critically important feature of scientific inquiry is that the unforeseeable tends to be of special significance just because of its unpredictability. *The more important the innovation, the less predictable it is, because its very unpredictability is a key component of importance*. Science forecasting is beset by a pervasive *normality bias*, because the really novel often seems so bizarre. A.N. Whitehead has wisely remarked: "If you have had your attention directed to the novelties in thought in your own lifetime, you will have observed that almost all really new ideas have a certain aspect of foolishness when they are first produced."[5] Before the event, a revolutionary scientific innovation will, if

5 Whitehead (1925, p. 49). Cited in Ziman (1978, p. 142).

imaginable at all, generally be deemed outlandishly wild speculation – mere science fiction, or perhaps just plain craziness.[6]

3 Is Future Knowledge Diminishing?

Some theorists have maintained that as science progresses the *magnitude of the issues* grows ever smaller and smaller. Later questions, so they hold, are always lesser questions, so that later science is always lesser science. Successive innovation becomes a matter of increasing refinement in detail, and furnishes new materials whose inherent significance decreases continually – exactly as with the decimal expansion of π or $\sqrt{2}$.

Scientific inquiry would thus be conceived of as analogous to terrestrial exploration, whose product – geography – yields results of continually smaller significance which fill in ever more minute gaps in our information. First explorers discover continents, later ones find river sources, and still later ones conquer mountains. On such a view, the later investigations yield findings of ever smaller scope and significance, with each successive accretion making a relatively smaller contribution to what has already come to hand. The advance of science leads, step by diminished step, toward a fixed and final view of things.

This general position is central to Charles Sanders Peirce's vision of ultimate convergence in scientific inquiry: "As the investigation goes on, additions to our knowledge...are of less and less worth. Thus, when Chemistry sprang into being, Dr. Wollaston, with a few test tubes and phials on a tea-tray, was able to make new discoveries of the greatest moment. In our day, a thousand chemists, with the most elaborate appliances, are not able to reach results which are comparable in interest with those early ones. All the sciences exhibit the same phenomenon" (Peirce 1934, sect. 7144).[7]

But such a theory encounters deep difficulties. For any such picture of convergence, however carefully crafted, will shatter against the *conceptual innovation* that continually brings entirely new, radically different scientific concepts to the fore and brings in its wake an ongoing wholesale revision of "established fact." Consider how many facts about a simple object – a sword, for example – were unknown to the ancients. They did not know that it contained carbon or that it conducted electricity. The very concepts at issue ("carbon,"

6 See Kuhn (1970), for an interesting development of the normal/revolutionary distinction.

7 See also Peirce's important 1898 paper on "Methods for Attaining Truth," in Peirce (1934, sects. 5.574 ff).

"electricity-conduction") were outside their cognitive range. There are key facts (or presumptive facts) even about the most familiar things – trees and animals, bricks and mortar – that were unknown a hundred years ago. This ignorance arises because the required *concepts* have not been formulated. It is not just that the scientists of antiquity did not know what the half-life of californium is but that they couldn't have *understood* this fact if someone had told them about it.

Ongoing scientific progress is emphatically not simply a matter of increasing accuracy by extending the numbers in our otherwise stable descriptions of nature out to a few more decimal places. It will not serve to take the preservationist stance that the old theories are generally acceptable as far as they go and merely need supplementation; significant scientific progress is genuinely revolutionary in that there is *a fundamental change of mind* as to how things happen in the world. Progress of this caliber is generally a matter not of adding further facts to tell ever diminishing gaps – on the order of filling in a crossword puzzle – but of changing the framework itself. The fact is that in natural science even small innovations with respect to data can engender large changes in theoretical systematization so as to render future science presently inscrutable.

4 Is Future Knowledge Perfectible?

How far can the scientific enterprise advance toward a definitive understanding of reality? Might science attain a point of recognizable completion?

But is the achievement of perfected science actually a genuine possibility, even in theory when all of the "merely practical" obstacles are put aside as somehow incidental? After all, what would *perfected science* be like? What sort of standards would it have to meet? Clearly, it would have to complete in full the discharge of natural science's mandate or mission.

It thus appears that if we are to claim that our science has attained a perfected condition, it would have to satisfy (at least) the four following conditions:

1. *Erotetic completeness*: It must answer, in principle at any rate, all those descriptive and explanatory questions that it itself countenances as legitimately raisable, and must accordingly explain everything it deems explicable.

2. *Predictive completeness*: It must provide the cognitive basis for accurately predicting those eventuations that are in principle predictable (that is, those which it itself recognizes as such).

3. *Pragmatic completeness*: It must provide the requisite cognitive means for doing whatever is feasible for beings like ourselves to do in the circumstances in which we labor.

4. *Temporal finality* (the omega-condition): It must leave no room for expecting further substantial changes that destabilize the existing state of scientific knowledge.

Each of these modes of substantive completeness deserves detailed consideration. First, however, one brief preliminary remark. It is clear that any condition of science that might qualify as "perfected" would have to meet certain formal requirements of systemic unity. If, for example, there are different routes to one and the same question (for instance, if both astronomy and geology can inform us about the age of the earth), then these answers will certainly have to be consistent. Perfected science will have to meet certain requirements of structural systematicity in the manner of its articulation: it must be coherent, consistent, consonant, uniform, harmonious, and so on. Such requirements represent purely formal cognitive demands upon the architectonic of articulation of a body of science that could lay any claim to perfection. Interesting and important though they are, we shall not, however, engage these *formal* requirements here, our present concern being with those four just-mentioned *substantive* issues.[8]

5 Theoretical Adequacy: Issues of Erotetic Completeness

Could we ever actually achieve erotetic completeness – the condition of being able to resolve, in principle, all of our (legitimately posable) questions about the world? Could we ever find ourselves in this position?[9]

In theory, yes. A body of science certainly could be such as to provide answers to all those questions it allows to arise. But just how meaningful would this mode of completeness be?

The reality is this erotetic completeness is an unattainable mirage. We can never exhaust the range of open questions.

For the world's furnishings are cognitively opaque; we cannot see to the bottom of them. Knowledge can become more extensive without thereby becoming more complete. And this view of the situation is rather supported than impeded if we abandon a cumulativist/preservationist view of knowledge or purported knowledge for the view that new discoveries need not supplement but can displace old ones.

8 Rescher's *Cognitive Systematization* (1979) deals with these matters.

9 Note that this is independent of the question "Would we ever want to do so?" Do we ever want to answer all those predictive questions about ourselves and our environment, or are we more comfortable in the condition in which "ignorance is bliss"?

It is sobering to realize that the erotetic completeness of a state of science *S* does not necessarily betoken its comprehensiveness or sufficiency. It might reflect the paucity of the range of questions we are prepared to contemplate – a deficiency of imagination, so to speak. When the range of our knowledge is sufficiently restricted, then its erotetic completeness will merely reflect this impoverishment rather than its intrinsic adequacy. Conceivably, if improbably science might reach a purely fortuitous equilibrium between problems and solutions. It could eventually be "completed" in the narrow erotetic sense – providing an answer to every question one can also in the then-existing (albeit still imperfect) state of knowledge – without thereby being completed in the larger sense of answering the questions that would arise if only one could probe nature just a bit more deeply. And so, our corpus of scientific knowledge could be erotetically complete and yet fundamentally inadequate. Thus, even if realized, this erotetic mode of completeness would not be particularly meaningful. (To be sure, this discussion proceeds at the level of supposition contrary to fact. The exfoliation of new questions from old in the course of scientific inquiry that is at issue in Kant's Principle of question-propagation spells the infeasibility of ever attaining erotetic completeness.)

After all, any judgment we can make about the laws of nature – any model we can contrive regarding how things work in the world – is a matter of theoretical triangulation from the data at our disposal. And we should never have unalloyed confidence in the definitiveness of our data base or in the adequacy of our exploitation of it. Observation can never settle decisively just what the laws of nature are. In principle, different law-systems can always yield the same observational output: as philosophers of science are wont to insist, observations *underdetermine* laws. To be sure, this worries working scientists less than philosophers, because they deploy powerful regulative principles – simplicity, economy, uniformity, homogeneity, and so on – to constrain uniqueness. But neither these principles themselves nor the uses to which they are put are unproblematic. No matter how comprehensive our data or how great our confidence in the inductions we base upon them, the potential reversatility of our claims cannot be dismissed.

We can reliably estimate the amount of gold or oil yet to be discovered, because we know the earth's extent and can thus establish a proportion between what we know and what we do not. But we cannot comparably estimate the amount of knowledge yet to be discovered, because we have and can have no way of relating what we know to what we do not. But (to hark back to Hegel), with respect to the realm of knowledge, we are not in a position to draw a line between what lies inside and what lies outside – seeing that, *ex hypothesi* we have no cognitive access to that latter. One cannot make a survey of the relative extent of knowledge or ignorance about nature except by basing it on some picture of nature that is already in hand – that is, unless one is prepared to take at face value the deliverances of existing science. This process of judging the adequacy of our science on its own telling is the best we can do, but it remains an essentially circular and consequently inconclusive way of proceeding. The long and short of it is that there is no cognitively adequate basis for maintaining the completeness of science in a rationally satisfactory way.

6 Pragmatic Completeness

The arbitrament of praxis over our scientific contentions – not then theoretical merit but practical applicability – affords the best standard of adequacy. But could we ever be in a position to claim that science has been completed on the basis of the success of its practical applications? On this basis, the perfection of science would have to manifest itself in the perfecting of control – in achieving a perfected technology. But just how are we to proceed here? Could our natural science achieve manifest perfection on the side of control over nature? Could it ever underwrite a recognizably perfected technology?

The issue of "control over nature" involves much more complexity than may appear on first view. For just how is this conception to be understood? Clearly, in terms of bending the course of events to our will, of attaining our ends within nature. But this involvement of "*our ends*" brings to light the prominence of our own contribution. For example, if we are inordinately modest in our demands (or very unimaginative), we may even achieve "complete control over nature" in the sense of being in a position to do *whatever we want* to do, but yet attain this happy condition in a way that betokens very little real capability.

One might, to be sure, involve the idea of omnipotence, and construe a "perfected" technology as one that would enable us to do literally *anything*. But this approach would at once run into the old difficulties already familiar to the medieval scholastics. They were faced with the challenge: "If God is omnipotent, can he annihilate himself (contra his nature as a *necessary* being), or can he do evil deeds (contra his nature as a *perfect* being), or can he make triangles have four angles (contrary to *their* definitive nature)?" Sensibly enough, the scholastics inclined to solve these difficulties by maintaining that an omnipotent God need not be in a position to do literally anything but rather simply anything that it *is possible* for him to do. Similarly, we cannot explicate the idea of technological omnipotence in terms of a capacity to produce any result whatsoever, wholly without qualification. We cannot ask for the production of a *perpetuum mobile*, for spaceships with "hyperdrive" enabling them to attain transluminar velocities, for devices that predict essentially stochastic processes such as the disintegrations of transuranic atoms, or for piston devices that enable us to set *independently* the values for the pressure, temperature, and volume of a body of gas. We cannot, in sum, ask of a "perfected" technology that it should enable us to do *anything* that we might take it into our heads to do, no matter how "unrealistic" this might be.

All that we can reasonably ask of it is that perfected technology should enable us to do anything *that it is possible for us to do* – and not just what we might *think* we can do but what we really and truly can do. A perfected technology would be one that enabled us to do anything that *can possibly* be done by creatures circumstanced as we are. But how can we deal with the pivotal conception of "can" that is at issue here? Clearly, only science – real, true, correct, *perfected* science – could tell us what indeed is realistically possible and what circumstances are indeed inescapable. Whenever our "knowledge" falls short of this, we may well "ask the impossible" by way of accomplishment (for example, spaceships in "hyperdrive"), and thus complain of incapacity to achieve control in ways that put unfair burdens on this conception.

Power is a matter of the "effecting of things possible" – of achieving control – and it is clearly cognitive state-of-the-art in science which, in teaching us about the limits of the possible, is itself the agent that must shape our conception of this issue. *Every* law of nature serves to set the boundary between what is genuinely possible and what is not, between what can be done and what cannot, between which questions we can properly ask and which we cannot. We cannot satisfactorily monitor the adequacy and completeness of our science by its ability to effect "all things possible," because science alone can inform us about what is possible. As science grows and develops, it poses new issues of power and control, reformulating and reshaping those demands whose realization represents "control over nature." For science itself brings new possibilities to light. (At a suitable stage, the idea of "splitting the atom" will no longer seem a contradiction in terms.) To see if a given state of technology meets the condition of perfection, we must *already* have a body of perfected science in hand to tell us what is indeed possible. To validate the claim that our technology is perfected, we need to *preestablish* the completeness of our science. The idea works in such a way that claims to perfected control can rest only on perfected science.

In the final analysis, then, we cannot regard the *realization* of "completed science" as a meaningful prospect – because we cannot really say science-independently what it is that we are asking for. And this consideration decisively substantiates the idea that we must always presume out knowledge to be incomplete in the domain of natural science.

7 Predictive Completeness

The difficulties encountered in using physical control as a standard of "perfection" in science will also hold with respect to *prediction*, which, after all, is simply a mode of *cognitive* control.

Suppose someone asks: "Are you really still going to persist in plaints regarding the incompleteness of scientific knowledge when science can predict *everything*?" The reply is simply that science will *never* be able to predict literally everything: the very idea of predicting *everything* is simply unworkable. For then, whenever we predict something, we would have to predict also the effects of making those predictions, and then the ramification of *those* predictions, and so on *ad indefinitum*. The very most that can be asked is that science put us into a position to predict, not *everything*, but rather *anything* that we might choose to be interested in and to inquire about. And here it must be recognized that our imaginative perception of the possibilities might be much too narrow. We can only make predictions about matters that lie, at least broadly speaking, within our cognitive horizons. Newton could not have predicted findings in quantum theory any more than he could have predicted the outcome of American presidential elections. One can only make predictions about what one is cognizant of, takes note of, deems worthy of consideration. In this regard, one can be myopic either by not noting or by losing sight of significant sectors of natural phenomena.

In the end it is science itself that determines the limits to predictability – insisting that some phenomena (the stochastic processes encountered in quantum physics, for example) are inherently unpredictable. And this is always to some degree problematic. After all, the most that science can reasonably be asked to do is to predict what is predictable. But this will have to be what it itself sees as in principle predictable. No more can be expected of science than answering every predictive question that it itself countenances as proper. And not only is this problematically circular, but we must once more recognize that any given state of science might have gotten matters quite wrong.

With regard to predictions, we are thus in the same position that obtains with regard to actually interventionist (rather than "merely cognitive") control. Here, too, we can unproblematically apply the idea of improvement – of *progress*. But it makes no sense to contemplate the achievement of *perfection*.

For its realization is something we could never establish by any practicable means.

8 Temporal Finality

And now on to temporal finality. Scientists from time to time indulge in eschatological musings and tell us that the scientific venture is approaching its end.[10] And it is, of course, entirely *conceivable* that natural science will come to a stop, and will do so not in consequence of a cessation of intelligent life but in C.S. Peirce's more interesting sense of completion of the project: of eventually reaching a condition after which even indefinitely ongoing inquiry will not – and indeed in the very nature of things *cannot* – produce any significant change, because inquiry has come to "the end of the road." The situation would be analogous to that envisaged in the apocryphal story in vogue during the middle 1800s regarding the Commissioner of the United States Patents who resigned his post because there was nothing left to invent (see Jeffrey 1940).

Such a position is in theory possible. But here, too, we can never effectively determine that it is actual.

There is no practicable way in which the claim that science has achieved temporal finality can be validated. The question "Is the current state of science, *S*, final?" is one for which we can never legitimate an affirmative answer. For the prospect of future changes of *S* can never be precluded. After all, one cannot plausibly move beyond "We have (in *S*) no good reason to think that *S* will ever change" to obtain "We have (in *S*) good reason to think that *S* will never change."

Moreover, just as the appearance of erotetic and pragmatic equilibrium can be a product of narrowness and weakness, so can temporal finality. We may think that science is unchangeable simply because we have been unable to change it. But that's just not good enough. Were science ever to come to a seeming stop, we could never be sure that it had done so not because it is at "the end of the road" but because we are at the end of our tether. We can never ascertain that science has attained the *X*-condition of final completion, since from our point of view the possibility of further change lying "just around the

10 This sentiment was abroad among physicists of the *fin de siècle* era of 1890–1900. (See Badash 1972, pp. 48–58.) And such sentiments are coming back into fashion today. See Feynmann (1965, p. 172). See also Stent (1969), and Hawkins (1981).

corner" can never be ruled out finally and decisively. No matter how final a position we *appear* to have reached, the prospects of its coming unstuck cannot be precluded. As we have seen, future science is inscrutable. We can never claim with assurance that the position we espouse is immune to change under the impact of further data – that the oscillations are dying out and we are approaching a final limit. In its very nature, science "in the limit" relates to what happens in the long run, and this is something about which we *in principle* cannot gather information: any information we can actually gather inevitably pertains to the short run and not the long run. We can never achieve adequate assurance that *apparent* definitiveness is *real*. We can never consolidate the claim that science has settled into a frozen, changeless pattern. The situation in natural science is such that our knowledge of nature must ever be presumed to be incomplete – and thereby inadequate overall.

One is thus led back to the stance of the idealistic tradition from Plato to Royce that human knowledge inevitably falls short of recognizably "perfected science" (the Idea, the Absolute), and must accordingly be looked upon as deficient. Our knowledge of the real is something we can certainly improve upon – but not something we can perfect. As best we can judge, science is destined to remain incomplete.

9 The Problem of Future Science in its Relation to Reality

How then are we to relate present-day science to the science of the future? The preceding considerations must inevitably constrain and condition our attitude toward the natural mechanisms envisaged in the science of the day. We certainly do not – or should not – want to reify (hypostasize) the "theoretical entities" of our *current* science as current science sees them – to say flatly and unqualifiedly that the contrivances of *our* present-day science correctly depict the furniture of the real world. We do not – or at any rate, given the realities of the case, should not – want to adopt categorically the ontological implications of scientific theorizing in just exactly the state-of-the-art configurations presently in hand. An unacceptable fallibilism precludes the claim that what we purport to be scientific knowledge is in fact *real* knowledge, and accordingly blocks the path to a scientific realism that maintains that the furnishings of the real world are exactly as our science states them to be inconclusive.

If the future is anything like the past, if historical experience affords any sort of guidance in these matters, then we know that *all* of our scientific theses and theories at the present scientific frontier will ultimately require revision in some (presently altogether indiscernible) details. All the experience we can muster indicates that there is no justification for viewing our science as more than an inherently imperfect stage within an ongoing development. The ineliminable prospect of far-reaching future changes of mind in scientific matters destroys any prospect of claiming that the world is as our science claims it to be – that present science's view of nature's constituents and laws is definitively correct.

Our prized "scientific knowledge" is no more than our "current best estimate" of the matter. The step of reification is always to be taken provisionally, subject to a mental reservation of presumptive revisability. We cannot but acknowledge the prospect that we shall ultimately recognize many or most of our frontier scientific theories to need revision and that what we proudly vaunt as scientific knowledge is a tissue of hypotheses – of tentatively adopted contentions of many or most of which we will ultimately come to regard as requiring serious revision or perhaps even abandonment.

And so a clear distinction must be maintained between "*our conception of* reality" and "reality *as it really is*." We must – and do – realize that there is precious little justification for holding that present-day natural science describes reality and depicts the world as it really is. And this constitutes a decisive impediment to any straightforward realism. It must inevitably constrain and condition our attitude towards the natural mechanisms envisioned in contemporary science. We certainly do not – or should not – want to reify (hypostatize) flat-out the "theoretical entities" of present-day science, to say flatly and without qualification that the contrivances of *our* present-day science correctly depicts the nature of things as they actually and ultimately are.

This situation blocks the option of scientific realism of any straightforward sort. Not only are we not in a position to claim that our knowledge of reality is *complete* (that we have gotten at the *whole* truth of things), but we are not even in a position to claim that our "knowledge" of reality is *correct* (that we have gotten at the *real* truth of things) in a way that pretends the need for any future qualification and ______dation. Such a position calls for the humbling view that just as we think our predecessors of a century ago had a fundamentally inadequate grasp on the "furniture of the world," so our successors of a millennium hence will take a similar view or our purported knowledge of things.

We do not – or at any rate, given the realities of the case, should not – want to adopt categorically the ontological implications of scientific theorizing in just exactly the state-of-the-art configuration presently in hand. A realistic acknowledgment of scientific fallibilism precludes the claim that the furnishings of the real world are exactly as our science states them to be – that electrons actually are just exactly as the latest *Handbook of Physics* claims them to be.

In relation to what is to come, our present thought about nature is no more than out inadequate anticipation – an estimate rather than a specification. And the fact of it is that we shall never be able to make claims about realty that go beyond what we presently think to be the case: reality as we can deal with it will always have to be *our* reality – reality as we presently conceive it to be.

10 Epistemic Vistas

Thought about nature is a complex issue: if only because nature's make-up can be known with different degrees of adequacy. The scale at issue in Display 1 is significant and instructive in this regard.

It seems unavoidable that when *X* is a large scale and complex fact regarding nature, then as one moves down that list one is soon out of one's depth, since in substantial matters of scientific inquiry we can seldom get beyond (2). And the fact of it is that we do and cannot but realize that there is a potential gap between (3) and (4) and that no matter when that future date may fall there is room for error.

Moreover, the instruction "Tell me what *X* is actually like, over and above and apart from what you think it to be" is an instruction that we cannot obey. The progressiveness of science means that the *merely seems/actually is* distinction is one we cannot operate with respect to matters of exact detail.

After all, who must *X* be in order for the following equivalence to hold:

Display 1
STEPS TOWARDS OBJECTIVITY

SUBJECTIVE	(1) I think that...	[SELF-IMPRESSION]
	(2) We (nowadays) think that...	[GROUP-IMPRESSION]
	(3) They will then think that...	[FUTURE IMPRESSION]
	(4) They will eventually or ultimately think that...	[LONG-RUN ULTIMACY]
OBJECTIVE	(5) It is to be thought (because actually true) that...	[IDEALITY]

X thinks *p* to be the case $\equiv$ *p* is actually the case

Pretty well nobody maintains that "X = the scientific community of the present day" will do the job. But for a time C.S. Peirce thought that the following would work out

X = the scientific community of the long-run future

But worries about the eventuation of the future led him to abandon this idea and move on to:

X = the scientific community in its *ideal* formation

And this step put Peirce into proximity with the later Josiah Royce who thought that

X = the Absolute (in its communal form)

would work out. But of course the problem is that we actually live in the real and not the ideal world. For those confined to the mundane idealistic of the actual, those idealizations do not afford much help.

It has to be acknowledged that with many matters – those of scientific futurity emphatically included – an assured knowledge of the truth (the whole truth and nothing but the truth) is inaccessible to finite minds. Those whose – like Euro-Idealists (German and British alike) – were to equate truth and knowledge have no alternative to following Josiah Royce into postulating an Absolute or Ideal Mind able to go where finite minds cannot reach. Granted, the idea of "naturalizing" the Absolute by reconceptualizing it as the work of finite minds in their aggregate totality is ingenious. But its visionary and unrealizable nature makes it an ultimately impracticable and futile expedient. We simply have to concede that Table I's ladder of objectivity cannot be descended as deep as we might like. The faith we lack in the present seems unavailable for the future also.

11 Science and Reality

We are now in a position to place into clearer relief one of the really big questions of philosophy: How close a relationship can we reasonably claim to exist between the answers we give to our factual questions at the level of scientific generality and precision and the reality they purport to depict?

Scientific realism is the doctrine that *science describes the real world*: that the world actually is as science takes it to be, and that its furnishings are as science envisages them to be.[11] But is such a "scientific realism" a tenable position?

It is quite clear that it is not. There is clearly insufficient warrant for and little plausibility to the claim that the world indeed is as our science claims it to be and that *our* science is *correct* science and offers the definitive "last word" on the issues. We really cannot reasonably suppose that science as it now stands affords the real truth as regards its creatures-of-theory.

One of the clearest lessons of the history of science is that where scientific knowledge is concerned, further discovery does not just *supplement* but generally *emends* the bearing of our prior information. Accordingly, we have little alternative but to take the humbling view that the incompleteness of our purported knowledge about the world entails its potential incorrectness as well. It is now a matter not simply of *gaps* in the structure of our knowledge, or errors of omission. There is no realistic alternative but to suppose that we face a situation of real *flaws* as well, of errors of commission. This aspect of the matter endows incompleteness with an import far graver than meets the eye on first view.[12]

Realism equates the paraphernalia of natural science with the domain of what actually exists. But this equation would work only if science, as it stands, has actually "got it right." And this is something we are surely not inclined – and certainly no *entitled* – to claim. We must recognize that the deliverances of science are bound to a methodology of theoretical triangulations from the data which binds them inseparably to the "state -of-the-art" of technological sophistication in data acquisition and handling.

As far as the actual condition of nature is concerned, science merely purports but does not achieve: it may provide us with the very best achievable estimate, but all the same this is still an estimate. The step of reification is always to be taken qualifiedly, subject to a mental reservation of presumptive revisability. We do and must recognize that we cannot blithely equate *our* frontier scientific theories with *the* truth. We do and must realize that the affirmations

---

11 For some classical discussions of scientific realism, see Sellars (1963); McKinnon (1972); Harré (1970); and Suppe (1977).

12 Some of the issues of this discussion are developed at greater length in Rescher's books *Methodological Pragmatism* (1977), *Scientific Progress* (1979), and *Cognitive Systematization* (1979).

of science are inherently fallible and that we can only "accept" provisionally subject to a clear realization that they may need to be corrected or even abandoned.

12 Ramifications of Impredictability

One further implication of these deliberations deserves comment. Man (*homo sapiens*) is an intelligent being whose actions in the world are guided by his beliefs – by what he takes to be knowledge in matters of fact. But insofar as the future development of human knowledge is impredictable, so this will also be the case with respect to human doings and dealings, so that human actions and affairs will also be in substantial measure impredictable. And thus insofar as man is an integral and interactive part of nature so there will be a sector of the natural world that will be impredictable as well. In relation to the cultivation of knowledge, the chance and contingency that afflicts human affairs is bound to involve some part of nature as well. To be sure, man and his dealings are small potatoes in the cosmic scheme of things. But nevertheless we are a part of the world's overall drama and here the role of our thought cannot simply be set at naught. And to this extent idealism, too, must be conceded its place.

References

Adams, M. 1987. *William Ockham*, vol. II. Notre Dame, Ind.: University of Notre Dame Press.

Anonymous. 1920. "The Future as Suggested by Developments of the Past Seventy-Five Years," in *Scientific American*, vol. 123, p. 321.

Badash, L. 1972. "The Completeness of Nineteenth-Century Science," in *Isis*, 63, pp. 48–58.

Feynmann, R. 1965. *The Character of Physical Law*. Cambridge, MA: The MIT Press.

Harré, R. 1970. *Principles of Scientific Thinking*. Chicago: The University of Chicago Press.

Hawkins, S.W. 1981. "Is the End in Sight for Theoretical Physics?," in *Physics Bulletin*, 32, 15–17.

Jeffrey, E. 1940. "Nothing Left to Invent," in *Journal of the Patent Office Society*, 22, 479–481.

Kuhn, Th. 1970. *The Structure of Scientific Revolution*, 2d ed. Chicago: The University of Chicago Press.

McKinnon, E. ed. 1972. *The Problem of Scientific Realism*. New York: Appleton.

Peirce, C.S. 1934. *Collected Papers*, Vol. V, edited by C. Hartshorne and P. Weiss. Cambridge, MA: Harvard University Press.

Rescher, N. 1977. *Methodological Pragmatism*. Oxford: Blackwell.

Rescher, N. 1979a. *Cognitive Systematization*. Oxford: Blackwell.

Rescher, N. 1979b. *Scientific Progress*. Oxford: Blackwell.

Rescher, N. 2000. *Kant and the Reach of Reason: Studies in Kant's Theory of Rational Systematization*. Cambridge: Cambridge. University Press.

Sellars, W. 1963. *Science Perception and Reality*. London: Humanities Press.

Stent, G. 1969. *The Coming of the Golden Age*. Garden City, NY: Natural History Press.

Suppe, F. ed. 1977. *The Structure of Scientific Theories*, 2nd ed. Urbana: University of Illinois Press.

Whitehead, A.N. 1925. *Science and the Modern World*. N. York: Macmillan.

Ziman, J. 1978. *Reliable Knowledge: An Exploration of the Grounds for Belief in Science*. Cambridge: Cambridge University Press.

The Limits of Future Knowledge: An Analysis of Nicholas Rescher's Epistemological Approach[1]

Amanda Guillan

Abstract

Regarding the limits of science, two different aspects should be initially considered. On the one hand, there are the "frontiers" or "barriers" (*Schranken*), which provide the criteria to separate science from non-science or science form pseudo-science. On the other, there are the "confines" (*Grenzen*), which are related to the problem of whether there is a ceiling of scientific knowledge that might be reached now or reached in the future. Nicholas Rescher usually focuses on the second aspect, although the problem of the "barriers" is also implicit in his approach, insofar as he sees science as a *system*.

Within this twofold presentation of the problem of the limits of science, this paper addresses the analysis of Rescher's proposal on the limits of scientific knowledge, and several steps are followed. First, there is an analysis of his approach to the "confines" of science, where he distinguishes limits in the weak sense (our current limitations) and limits in the strong sense (the unsolvable questions). Second, the attention goes to the implications of this approach for the problem of the limits as "barriers," which are related to the idea of a system. Third, the connections of the limits of science (both as "confines" and as "barriers") with the issue of scientific progress are considered. In this case, the focus is on "conceptual creativity," which the paper analyses from the concept of historicity.

Keywords

Limits – future – knowledge – Rescher – epistemology

When the limits of science are considered from an epistemological angle, there are at least two different aspects that can be taken into account: borders and ceiling. (i) The problem of the existence of the limits of science as "frontiers" or "barriers" (*Schranken*), that is, what are the boundaries that

1 This research is supported by the *Program FPU* of the Spanish Ministry of Education, Culture and Sport. A previous version of this paper was delivered at the *Conference on the Limits of Science* (*Jornadas sobre los Límites de la Ciencia*) held at the University of A Coruña, Ferrol Campus, on 14–15 of March 2013.

 DOI 10.1163/9789004325401_008

separate science from non-science. (ii) The issue of the possibility of the limits from the perspective of the "confines" (*Grenzen*), i.e., the question of the possible ceiling of scientific knowledge that can be reached by scientific activity (Radnitzky 1978).

In his philosophical analysis of the limits of science, Nicholas Rescher emphasizes the epistemological angle of the discussion. In this regard, he has paid special attention to the second aspect: the "confines" or "ceiling" of the knowledge involved in the scientific activity. This question is related to the imperfect nature of science and the unpredictability of future knowledge.[2] However, insofar as he sees science as a system, in his conception is also implicit that there are "frontiers" that separate science from other human activities.

Within this framework, this chapter analyses Rescher's proposal about the limits of scientific knowledge, which is certainly related with his conception of science as a system. To do this task, several steps are followed. First, the attention goes to his approach to the limits of science as "confines." In this regard, two kind of limits are considered: (a) the limits in the weak sense (the current difficulties in solving a particular problem); and (b) the limits in the strong sense (the questions that science will never be able to solve).

Second, the focus is on the implications of this approach. Thus, the limitations of science as a system are also analyzed.[3] This involves, on the one hand, the consideration of the limits as "barriers" and, on the other hand, the analysis of the implications that Rescher's approach has for scientific progress. In this regard, the role of "conceptual creativity" is emphasized.[4] From this perspective, the notion of *historicity* seems to have a major role, both with regard the limits of science (as "confines" and as "barriers") and in relation to conceptual change.

1 The Limits of Science as "Confines" and Scientific Progress

Regarding the limits of science, Rescher's epistemological focus is commonly on the issue of the "confines," because in his approach scientific knowledge of the future is important, and this is related to the limits of future knowledge. Thus, he analyzes the obstacles that can affect the attainable knowledge by scientific research in the future, which include the difficulties in predicting these obstacles.

2 Cf. Rescher (1984). Later, a revised edition of this book was published: Rescher (1999b).

3 On his conception of science as a system, see Rescher (1992; especially, Sec. 2, pp. 61–125).

4 On scientific creativity and its relations to technological innovation, see Gonzalez (2013).

Moreover, Rescher has noted that there is a difference between two different meanings of "limits" related to science: limits *in the weak sense* and limits *in the strong sense*. (a) There are limits *in the weak sense* when we cannot answer questions *now*, because the available knowledge is not enough – or is not good enough – to solve them. Meanwhile, (b) there are limits *in the strong sense* when there are questions that we will *never* be able to answer, even in a very distant future (Rescher 2001a, pp. 73–75).

1.1 *Limits in the Weak Sense: The Imperfect Nature of Science*

In Rescher's conception, science is clearly subject to limits in the weak sense. He understands by "limits in the weak sense" those current difficulties in solving a particular problem. So they are limits that can be overcome in the future, when the needed knowledge becomes available or when the research means (economic, technological, etc.) increase or improve. Nevertheless, it is a kind of limitation that will always concern science, since there are questions for which it is impossible to find an answer at each stage of scientific development. This is due to the imperfect nature of science, which in fact is *our* science (Rescher 2006a).

According to Rescher, science should possess four conditions in order to be considered a perfected science: (1) erotetic completeness, (2) predictive completeness, (3) pragmatic completeness, and (4) temporal finality (the omega condition). He thinks that the three first ones are problematic from a theoretic point of view and also unfeasibly from a pragmatic perspective. At the same time, he thinks that the temporal finality is unattainable, due to the internal link between scientific progress and technological innovation (Rescher 1999a, pp. 137–148).

The *erotetic completeness*, which is the first requirement of a perfected science, entails that science must answer all descriptive and explanatory questions of its domain. That is, in order to reach erotetic completeness, science must "explain everything it deems explicable" (Rescher 1999b, p. 147). Here, there are two issues that make erotetic completeness impossible. On the one hand, the dynamics of scientific research that is developed in a question-answer process; and, on the other, the Kantian "principle of question propagation," which has a general character and affects science.

Certainly, scientific progress is linked to the ability of science to solve questions. But, as Rescher notes, we can only ask those questions that are already implicit on the available knowledge (Rescher 1999b, p. 6). This means that science is subject to limits, to the extent that there will be always important questions that we cannot ask until those limits were overcome. We also have to keep in mind the Kantian *principle of question propagation*, which involves that "the answer given to any empiric (scientific) question cause subsequent questions that must to be solved" (Kant 1783/1968, p. 352; quoted in Rescher 1999a, p. 120). Undoubtedly, this makes it impossible to achieve the erotetic completeness.

With regard to *predictive completeness* – the second condition of a perfected science – a thoroughly complete science must predict accurately everything it deems, in principle, predictable. However, Rescher's epistemological perspective is a falibilistic one, both concerning scientific knowledge, in general, and with respect to predictive knowledge, in particular. Thus, he thinks that any knowledge that we accept is open to revision because it can be false (Rescher 1999b, Ch. 3, pp. 29–42). In addition, since prediction is a specific kind of knowledge about the future, Rescher considers a series of specific obstacles to scientific prediction. They are principally epistemological and ontological limits. Among them, uncertainty, chance and chaos can be emphasized (Rescher 1998, Ch. 8, pp. 133–156; see also Gonzalez 2012a).

Pragmatic completeness, which is the third requirement every science must have in order to be seen as finished or completed science, is also problematic. To achieve this goal, Rescher states that science "must provide the requisite cognitive means for doing whatever is feasible for beings like ourselves to do in the circumstances in which we labor" (Rescher 1999b, p. 146). Science must provide us a perfect control, and this implies that we must obtain a perfect technology too (at least for a series of sciences, such as natural sciences).

Rescher has pointed out two conclusions that can be drawn about the possible perfection of technology: (i) there are no reasons to believe that it can be feasible; and (ii) even if we reach technological perfection, we can't evaluate it (Rescher 2012). In this regard, the limits that affect technology should be considered. They are related to the problem of complexity involved by technological innovation. So the limits of ability – due to physical possibility of doing something – and limits of management – that are derived from complexity management – bring us to the conclusion that it is impossible to reach the pragmatic perfection (Rescher 1999b, pp. 166–170).

Temporal finality – the fourth requirement, which is also called "the omega condition" by Rescher (2003, p. 259) – is the feature that, in his judgment, characterizes a perfect science. To accomplish temporal finality, the scientific knowledge achieved in a concrete moment must not allow future changes. Rescher maintains that this condition is unrealizable, because of the Kantian principle of question propagation. Also, when he considers the link between scientific progress – mainly in natural sciences – and technological innovation, he thinks that temporal finality does not seem feasible either (Rescher 1999a, p. 146).

In this regard, it should be emphasized that Rescher describes science and technology as "two legs of the same body" (1999a, p. 100). In such a case, although they are distinguishable activities, they are mutually required, so each one of them impels the development of the other. This involves that science needs the tools provided by technology, and at the same time technology requires knowledge produced by science. Consequently, it happens that technological innovations contribute to scientific development, and scientific developments result in technological innovation too. Therefore also *technological limits* hinder scientific progress.

Thus, Rescher sees the limits *in the weak sense* in connection with the imperfect nature of science. This can be seen in several consecutive levels: conceptual limits, limits of prediction, limits of technology, and economic limits. When the limits of science are investigated, it seems clear that Rescher gives priority to the "internal" side of limits (the limits of scientific activity itself) over the "external" aspect (that is related to the obstacles that arise from the relation between science and its context). At the same time, his contribution about the limits highlights that he is thinking about natural sciences, in general, and physics, in particular.

(1) There are *conceptual* limits concerning our ability to ask questions. Thus, we might think that there are important questions in current science that fall outside our conceptual horizons. This affects the explanatory ability of science, and it also has a negative effect on scientific prediction, insofar as we cannot predict those things we cannot conceive (Rescher 2012). (2) There are specific limits to *prediction* too. These limits prevent science from achieving predictive completeness. In this regard, Rescher has pointed out a series of epistemological and ontological factors that are obstacles to scientific prediction.[5]

(3) Furthermore, the development of science is hindered by the limits of *technology*. In the case of the natural sciences, technological tools are especially important for observation and experimentation. So the limits of technology – that Rescher sees in relation to complexity – can be also limits for science. (4) There are also *economic* limits to scientific research. They can be clearly seen when the interdependence between scientific progress and technological innovation is highlighted. In fact, Rescher considers that economic limits are especially relevant. In his judgment, the limits of science "are fundamentally economic limits imposed by the technological character of our access to the phenomena of nature" (Rescher 2003, p. 251).

Undoubtedly, the aforementioned limits make it impossible that science becomes completed. Furthermore, the limits of science should be considered

---

5 On Rescher's proposal concerning the limits of prediction, cf. Guillan (2012). On his approach to explanation and prediction, see van Fraassen (2008).

with regard to the historic development of science, insofar as they are *current* limits, i.e., they are not intrinsic limitations, so they can change. From this perspective, the issue of the limits can be considered with regard to *historicity*, which is related to three successive levels: science, agents and the reality itself researched.[6] In my judgment, historicity can contribute to open a new perspective for the research into the limits of science. Moreover, it allows us to broaden Rescher's proposal that usually gives priority to "internal" limitations (the limits of scientific activity) over "external" obstacles (concerning agents and reality itself researched).

1.2 *The Strong Sense of "Limits:" Unsolvable Problems* (Insolubilia) *and the Unpredictability of Future Science*

Rescher thinks that the strong sense of limits is more problematic than the limits considered in the weak sense. In his view, limits in the strong sense are equated with the existence of unsolvable scientific problems (*insolubilia*). In this regard, when we accept that science has limitations in the strong sense, it is accepted that there are now (or there will be in the future) significant questions that will never be solved. They are questions whose solution lies beyond the limits of science, even in the long term.

However, Rescher notes that "there is no reason to think, on the basis of general principles, that any issues within the domain of natural science lie beyond its capabilities" (Rescher 1999b, p. 3). From this perspective, epistemological limits that affect science are only limits in the weak sense. In such a case, this stance – that science is bound by limits in the weak sense – is compatible with the thesis that any question raised at certain point of scientific inquiry can be solved, at least in the future.

But, on the basis of *unpredictability* of future science, it cannot be claimed with certainty that there will be an answer to every question raised by science.[7] So Rescher, in principle, accepts the possibility that there are epistemological limits in the strong sense, that entail that it is possible to find unsolvable problems in scientific inquiry. Nevertheless, he notes that the claim that there may be unsolvable problems is different from the assertion that these unsolvable

6 On historicity as a trait of science in the three aforementioned levels, and on his influence on conceptual change, cf. Gonzalez (2011b).

7 There is a crucial distinction between "unpredictable" and "not-predictable." "Unpredictable" means the complete impossibility of prediction. It is fundamentally due to the presence of phenomena characterized by the anarchy or the lack of laws. By contrast, "not-predictable" refers to current impossibility of predicting, normally due to phenomena instability. This distinction is in Gonzalez (2010, p. 289).

problems can be identified. Rescher calls this second assertion – according to which we can identify now the questions that science will never solve – "hyperlimitation" (Rescher 1999b, p. 113).

"Hyperlimitation" is a matter that must be seen in relation with problems for predicting future knowledge. As Rescher sees this issue, it should be admitted that there are difficulties in identifying "unsolvable" scientific problems (*insolubilia*) (Rescher 1998, pp. 186–188; and 1999b, Ch. 8, pp. 111–127); because there are also difficulties in predicting how future science will be (Rescher 1998, pp. 177–183). In fact, he notes that "the prospect of present knowledge about future discoveries is deeply problematic since the future of knowledge is fundamentally unpredictable. The details of the cognitive future are hidden in an impenetrable fog" (Rescher 2001a, p. 64).

This issue is related to the Kantian principle of question propagation. In Rescher's epistemological judgment, this principle has two fundamental consequences for science: (i) the unpredictability of future science, and (ii) the impossibility of achieving a perfect science. On the basis of current knowledge, the Kantian principle of question propagation involves the impossibility of predicting what questions will be asked in the future, and certainly we do not know now what the answers to these questions will be. Therefore, unsolvable problems cannot be identified and its existence as such unsolvable problems cannot be determined (that is, not as mere current difficulties that can be overcome in the future).

Clearly, the advancement of science can be seen as a broad cognitive process of "Kantian inspiration." From this perspective, scientific advancement is a question-answer process, where every new answer has repercussions on the very questions that can be raised. In this regard, Rescher points out that new knowledge can affect the questions we ask in three different ways: (a) it can give new answers to old questions; (b) it can generate new questions; and (c) it can led us to discover the illegitimacy of some old questions (Rescher 1999b, p. 12).

Consequently, he considers that we cannot predict accurately the content of future knowledge. Even more, we cannot foresee what questions we will ask in the future, and we cannot say with certainty that every current unsolved question is a legitimate question. Thus, "the task of specifying the limits of scientific capability in the production of knowledge is itself one that transcends the limits of our cognitive powers" (Rescher 2009. p. 18). In this way, identifying unsolvable scientific problems is a task that lies beyond our abilities, even when we can assume that these problems exist.

We might notice that Kantian principle of question propagation means the impossibility of scientific completeness. Rescher connects this with the existence of limits in the weak sense. So the complete opening to scientific knowledge in the future leads him to dismiss the ideal of perfected science; that is, the possibility of a science thoroughly finished (in the sense of fully accomplished) [Rescher 1999b, pp. 145–176], because every scientific advancement will produce new questions that need an answer, and so on. From this viewpoint, science is developed in a question-answer process, as long as it can overcome the limits that hinder its advancement.

As Rescher sees this issue, the distinction between limits *in the weak sense* and limits *in the strong sense* – where the case of *hyperlimitation* can be included – has major implications for scientific progress. His proposal is clear: "The distinction between these various types of limits thus carries the important lesson – already drawn by Kant – that even the resolution of all *our* scientific problems would not necessarily mean that science as such is finite or completable" (Rescher 2001a, p. 74). In this way, Rescher maintains a view of scientific progress as unlimited (i.e., always open to the future) [Rescher 1978].

On the one hand, he sees science as subject to limits *in the weak sense*. They are current limits that have been present at every historical moment of scientific development, and that will affect science also in the future. But they are obstacles that can be overcome by scientific development itself. Rescher maintains that "the essential limitlessness of science on the side of terminating limits – the feasibility of unending scientific progress – is not to deny the prospect of problems whose solution lies beyond the physical and/or economic limits of man's investigative capacities. The existence of actually unanswerable questions in science – problems whose solution lies forever on the inaccessible side of a technologically imposed data-barrier – would *not* mean an eventual end to scientific progress" (Rescher 2006a, p. 73).

To sum up, Rescher considers that always there will be limits to scientific progress. But, at the same time, he thinks that scientific progress is always possible, insofar as these limits can be overcome. Concurrently, he rejects the existence of limits *in the strong sense*. First, we may doubt if those things that science considers infeasible are, in fact, part of the scientific domain; and, second, even if these limitations exist, it is not possible for us to identify them (Rescher 2001a, p. 75). Therefore, although we should admit the presence of limits for scientific progress, it is questionable whether there would be a point where scientific advancement stops.

2 The Limits of Future Knowledge in a Systematic Conception of Science

This approach of Rescher to the limits of science as "confines" should be related with his systematic conception of science. This allows us to address the distinction between the *internal* limits of science, where the content of knowledge has a key role, and the *external* limits of scientific activity, where its knowledge requires social acceptance. At the same time, the view of science as a system involves the consideration of the limits as "frontiers" or "barriers," insofar as the elements within the system as a whole can be distinguishable, even though they are interdependent in relevant ways.

Since a system is, in principle, something that is closed, there must be barriers that separate it from other different realms. Nevertheless, the limited nature of science is compatible – in Rescher's approach – with an unlimited view of scientific progress (where *conceptual creativity* plays a major role). Thus, he sees science as a system that is also unlimited on the basis of conceptual creativity, but he hardly pays attention to the historicity that modulates scientific change, also from the viewpoint of concepts.

2.1 *Limitations of Science as a System*

The concept of *system* has a clear presence in the large philosophical production of Rescher, in general, and in his epistemology, in particular. From the perspective of a system, the distinction between the "internal" obstacles – those related to scientific activity itself –, and the "external" obstacles, which come from the environment, is highlighted. He usually focuses his attention on the internal obstacles to scientific activity, a matter he analyses according to the distinction between limits *in the weak sense* and limits *in the strong sense*. In this regard, there are internal limits to science rooted in the constitutive elements of science, so they can be found in its language, structure, knowledge, method, activity, aims, and values (among them, ethical ones) [Gonzalez 2010, pp. 275–276].[8]

But there are external limits of science too. The external limits arise from the relation between science and its environment (natural, social or artificial). These limits are related to the complexity that hinders scientific knowledge and that – in Rescher's approach – especially affects prediction.[9] So when the future that we want to predict is developmentally open it can become unpredictable. This happens with future knowledge, which is inaccessible to our current knowledge categories (above all when it is a future in the long term and we want a detailed prediction) [Rescher 2012].

Within the framework of his idea of system, Rescher addresses the problem of the limits as "barriers." In this regard, he notes that science is a limited

8 On constitutive components of science, cf. Gonzalez (2005, pp. 3–49; especially, pp. 10–11). See also its connection with the distinction between basic science, applied science, and application of science in Gonzalez (2013, pp. 17–18).

9 On complexity in science, see the papers in Gonzalez (2012b). On how complexity affects prediction, especially in the case of economics, see Gonzalez (2011a).

enterprise because scientific knowledge is only a *human good* among others (Rescher 1999a, pp. 103–105). He considers other ways of knowing, such as philosophy and humanistic disciplines. However, scientific knowledge is an especially important good, because it also has an instrumental value. From a pragmatic viewpoint, looking for knowledge as a good "impedes in no way the cultivation of other legitimate goods. On the contrary, it helps – and makes it easier – to pursuit them, achieving thus an instrumental value that it is added to its value as an absolute good in its own right" (Rescher 1999a, p. 105).

Although scientific knowledge is only one of the many ways of knowing, for Rescher it is "sovereign" in its realm (1999a, p. 116). Thus, he insists on autonomous nature of science (1999a, pp. 114–116). Autonomy of science involves that scientific outcomes cannot be evaluated from rationality external to science. This affects the self-corrective nature of scientific knowledge: scientific theories' faults only can be corrected by other theories (and, therefore, by scientists). Consequently, although science is only a cognitive enterprise among others, he considers that it is fully self-sufficient with regard to the explanation and prediction of phenomena (natural, social or artificial).

Thus, in Rescher's conception, science is an autonomous and limited system in the two aforementioned senses: (a) while there are "frontiers" or "barriers" that separate science from non-science (and also science form pseudo-science); and (b) regarding the "confines" or possible "ceiling" for scientific activity. These can be "confines" in the weak sense or in the strong sense. Since there are problems for identifying questions as examples of limits in the strong sense, he thinks that the analysis of these limits is problematic. In fact, he questions the scientific nature of unsolvable issues, because we can only maintain that certain matter lies in principle beyond the limits of science "and this would gravely compromise the legitimacy of the question as a genuinely scientific one" (Rescher 2001a, p. 75).

It happens that, in Rescher's approach, science is *our* science (Rescher 1992, pp. 110–125). Thus, an unsolvable problem is so in relation to human abilities, conceptual system, or the means at our reach, that can always improve. From this viewpoint, there is no reason to think, in principle, that current unsolvable problems cannot be answered in the future. Therefore, Rescher wants to harmonize his proposal about the limits of science with a conception of progress as potentially unlimited. In order to do that, he highlights the process nature of scientific activity.

In his judgment, "science, properly understood, is not a body of theses or theories, but a process – an ongoing venture in inquiry whose products are ever changing" (Rescher 1996, p. 139). (i) For Rescher, these changes are fundamentally conceptual ones, so we cannot anticipate them from our current knowledge categories. (ii) The unpredictability of future knowledge implies the impossibility of determining the existence of limits in the strong sense. (iii) This proposal also involves a falibilistic approach of scientific knowledge, to the extent that the presence of error cannot be ruled out. This highlights that it is always possible for science to improve.

2.2 *Conceptual Creativity from the Viewpoint of Historicity*

Undoubtedly, knowledge in science is related to *process*, which is notion that has a key one in Rescher's conception.[10] It allows him to join an approach of cognitive progress as potentially unlimited as well as a view of science where there are limits. This matter connects with *conceptual creativity*, which he relates to scientific progress and the limited nature of science. In his words, "ongoing progress in scientific inquiry is a process of *conceptual* innovation that always places certain developments outside the cognitive horizons of earlier workers because the very concepts operative in their characterization become available only in the course of scientific discovery itself" (Rescher 2012, p. 151).

In my judgment, an appropriate characterization of conceptual creativity as the decisive factor of scientific progress should take into account the *historical dimension* involved in scientific change. This is also noted by Rescher when he points out that "our conceptions of things always present a moving rather than a fixed object of scrutinity, and this historical dimension must also be reckoned with" (Rescher 1992, p. 246). However, in despite of this admission of the historical dimension, he does not use to consider the historicity of human knowledge, in general, and scientific knowledge, in particular. Specifically, in spite of his objections to P.F. Strawson when the processes are considered, it is required a greater presence of historicity.[11]

Rescher criticizes that Strawson gave conceptual priority to substances over processes (Rescher 1996, pp. 60–64). Thus, he defends a "conceptual processism," according to which "processes are conceptually fundamental *vis à vis* things in that (1) the explanatory characterizations of what a thing is always involves recourse to a processual account of what that thing does; and (2) the identification of any particular thing as such always involves reference to various processes, a thing being constituted as what it is by means of identificating processes" (Rescher 1996, p. 58).

10 As a matter of fact, Rescher has a clear interest in the metaphysical processes, which can be related to the Leibnizian roots of his philosophy. See, in this regard, Rescher (1996).

11 On the role played by historicity in the understanding of scientific change, cf. Gonzalez (2008 and 2011b).

He suggests that it is necessary – in order to know phenomena – to organize the whole set of concepts and ideas that provide us with the framework required for accomplish this task. On the ontological basis, we find that "processes rather than things best represent the phenomena that we encounter in the natural world about us" (Rescher 1996, p. 2). Future knowledge goes hand in hand with his defense of process metaphysics in the face of a substantialist approach. In fact, he considers that substantialism has problems for address the future form an ontological perspective, since future phenomena have not occurred yet. This problem disappears when processual nature of phenomena is highlighted, since "the processual nature of the real means that the present constitution of things always projects beyond itself into one as yet unrealized future" (Rescher 1996, p. 54).

However, Rescher's references to the historicity of knowledge are scarce, and – in my judgment – his contributions in this respect are overly general. Although he expressly recognizes that "our science is a historical phenomenon: it is one transitory state of things in an ongoing process" (Rescher 1999b, p. 38), he has not developed how historicity modulates the concepts that articulate knowledge. In fact, historicity does not appear in his criticism on Strawson. This is a key problem in Strawson's approach to concepts because, although he highlights that concepts are more than mere mental representations, he fails in "the acknowledgment of ontological character of historicity because the processes of things and agents refers to a reality that is changeable" (Gonzalez 2008, p. 74).

It is important to note that *historicity* – which involves the realm of reality – is compatible with the objectivity of scientific knowledge (Gonzalez 2008, pp. 68–71). If we accept historicity as a feature of human activity, it must be incorporated into scientific concepts. Thus, concepts change in order to get changes happened in the reality. Wenceslao J. Gonzalez points out this: "accepting the objectivity of concepts involves incorporating the historicity of science, agents, and reality itself (above all, social and artificial)" (Gonzalez 2008, p. 68).

In my judgment, science is open to conceptual innovation. This is a key feature of scientific progress and the limits of science, since scientific research is always accomplished from a conceptual framework that is open to changes. Indeed, conceptual change can contribute to overcome our current limitations, while every particular conceptual framework limits scientific research. Rescher admits that "the major discoveries of later stages are ones which the workers of a substantially earlier period (however clever) not only have failed to make but which they could not even have understood, because the requisite concepts were simply not available to them" (Rescher 2012, p. 151).

This component of variability has an effect on the problem of the limits of science. In fact, *historicity* concerns three successive levels of analysis: science, agents and reality itself researched. This is highlighted by Wenceslao J. Gonzalez: "historicity (*Geschichtlichkeit*/*historicidad*) is a trait of science, in general, and each science, in particular. This facet can be found in the whole set of constitutive elements of science, such as language, structure, knowledge, method, activity, ends, and values. (2) Historicity configures the agents themselves involved in the development of scientific research, insofar as they are human beings within a historical context. (3) Historicity is a characteristic of the reality itself that is researched (above all, in the social and artificial realms)" (Gonzalez 2011b, p. 43).

From this perspective, historicity can also modulate the limits of science in the three levels aforementioned. Firstly, there are limits to activity displayed by science, both in the general level and in each science in particular. These are limits that concern constitutive elements of science, so they can be found in scientific language, structure, knowledge, method, activity, aims and values. Secondly, there are also limits due to the researchers themselves as human agents that are placed in a particular historical context. Thirdly, limits of science can be so according to the reality itself that is researched (natural, social or artificial).

3 Final Remarks

Regarding the limits of science from an epistemological viewpoint, Rescher proposes an approach that combines Kantian elements with pragmatist proposals (Rescher 2001b). He also takes into account the different levels of analysis. Thus, within the distinction between limits as "frontiers" or "barriers" and limits as "confines" or "ceiling," his approach on the "barriers" is placed in the framework of a systemic conception of science. Indeed, as a *system*, science has some frontiers that separate it from other human activities.

Later, in his proposal on the epistemological limits, there are two different aspects when the "confines" of science are investigated. (1) The limits appear in the weak sense, that is, as limitations that hinders scientific progress in a particular historical period; but these limitations can be overcome in the future. (2) There are limits in the strong sense that are related to unsolvable problems in any historical stage of science. Limits in the weak sense have to do with the imperfect nature of science. They are current limits rather that intrinsic limitations, so they can change in the future.

Therefore, the limits must be considered in connection with historical development. Additionally, they influence an approach to scientific progress as *unlimited*, because every time that a limit is overcome, new limitations appear, and we have to overcome them. Rescher also questions the existence of limits in the strong sense. He relates this matter to the difficulties in predicting future knowledge because, in his judgment, these difficulties involve the impossibility of identifying unsolvable problems. In this way, identifying unsolvable problems is itself a task outside the limits of what is possible to us.

When he considers the problem of the limits of science – both in the sense of "barriers" and from the perspective of "confines" – his proposal must be placed in the framework of a systematic conception of science where the notion of *process* is a key one to understand scientific change, in general, and conceptual change, in particular. However, he does not use to consider the *historicity* of science, although he is interested in the changes produced in science by means of conceptual creativity.

I think that a suitable characterization of the limits of science must take into account historicity as a feature concerning three successive levels of analysis: science, agents and the reality researched (Gonzalez 2011b). In such a case, (a) it can be considered the limits in the activity itself displayed by science, so that there are limits which affects language, structure, knowledge, method, activity, ends and values of science. (b) Also the limits due to agents can be considered, insofar as they are human beings placed in a particular historical context. (c) Moreover, there are limits concerning the reality itself researched (natural, social or artificial). Thus, if we consider the historicity that modulates scientific change in these three levels (science, agents and reality itself researched), it is possible – in my judgment – to offer a wider analysis about the problem of the limits of science.

References

Gonzalez, W.J. 2005. "The Philosophical Approach to Science, Technology and Society," in Gonzalez, W.J. ed. *Science, Technology and Society: A Philosophical Perspective*, (pp. 3–49). A Coruña: Netbiblo.

Gonzalez, W.J. 2008. "El enfoque cognitivo en la Ciencia y el problema de la historicidad: Caracterización desde los conceptos," in *Letras*, 79 (114), 51–80.

Gonzalez, W.J. 2010. *La predicción científica. Concepciones filosófico-metodológicas desde H. Reichenbach a N. Rescher*. Barcelona: Montesinos.

Gonzalez, W.J. 2011a. "Complexity in Economics and Prediction: The Role of Parsimonious Factors," in Dieks, D., Gonzalez, W.J., Hartman, S., Uebel, Th. and Weber, M. eds. *Explanation, Prediction, and Confirmation*, (pp. 319–330). Dordrecht: Springer.

Gonzalez, W.J. 2011b. "Conceptual Changes and Scientific Diversity: The Role of Historicity," in Gonzalez, W.J. ed. *Conceptual Revolutions: From Cognitive Science to Medicine*, (pp. 39–62). A Coruña: Netbiblo.

Gonzalez, W.J. 2012a. "Methodological Universalism in Science and its Limits: Imperialism versus Complexity," in Brzechczyn, K. and Paprzycka, K. eds. *Thinking about Provincialism in Thinking*, (pp. 155–175). Poznan Studies in the Philosophy of the Sciences and the Humanities, vol. 100. Amsterdam/New York: Rodopi.

Gonzalez, W.J. ed. 2012b. *Las Ciencias de la Complejidad: Vertiente dinámica de las Ciencias de Diseño y sobriedad de factores*. A Coruña: Netbiblo.

Gonzalez, W.J. 2013. "The Roles of Scientific Creativity and Technological Innovation in the Context of Complexity of Science," in Gonzalez, W.J. ed. *Creativity, Innovation, and Complexity in Science*, (pp. 11–40). A Coruña: Netbiblo.

Gonzalez, W.J. 2016. "Rethinking the Limits of Science: From the Difficulties for the Frontiers to the Concern on the Confines," in Gonzalez, W.J. ed. *The Limits of Science: An Analysis from "Barriers" to "Confines."* Brill, pp. 3–30.

Guillan, A. 2012. "Límites del conocimiento y Ciencias de la Complejidad: Factores epistemológicos y ontológicos como obstáculos a la predicción científica," in Gonzalez, W.J. ed. *Las Ciencias de la Complejidad: Vertiente dinámica de las Ciencias de Diseño y sobriedad de factores*, (pp. 181–204). A Coruña: Netbiblo.

Kant, I. 1783/1968. *Prolegomena zu einer künftigen Metaphysik*. English version in Kant, I. *Werke*, vol. 4. Berlin: Walter de Gruyter.

Radnitzky, G. 1978. "The Boundaries of Science and Technology," in *The Search for Absolute Values in a Changing World. Proceedings of the VIth International Conference on the Unity of Sciences*, vol. II, (pp. 1007–1036). New York: International Cultural Foundation Press.

Rescher, N. 1978. *Scientific Progress. A Philosophical Essay on the Economics of the Natural Science*. Oxford: B. Blackwell.

Rescher, N. 1984. *The Limits of Science*. Berkeley: University of California Press.

Rescher, N. 1992. *A System of Pragmatic Idealism. Vol. I: Human Knowledge in Idealistic Perspective*. Princeton, NJ: Princeton University Press.

Rescher, N. 1996. *Process Metaphysics. An Introduction to Process Philosophy*. Albany, NY: State University of New York Press.

Rescher, N. 1998. *Predicting the Future. An Introduction to Theory of Forecasting*. Albany, NY: State University of New York Press.

Rescher, N. 1999a. *Razón y valores en la Era científico-tecnológica*. Barcelona: Paidós.

Rescher, N. 1999b. *The Limits of Science*, revised edition. Pittsburgh, PA: University of Pittsburgh Press.

Rescher, N. 2001a. "On Learned Ignorance and the Limits of Knowledge," in Rescher, N. *Cognitive Pragmatism. The Theory of Knowledge in Cognitive Perspective*, (pp. 63–80). Pittsburgh, PA: University of Pittsburgh Press.

Rescher, N. 2001b. *Cognitive Pragmatism. The Theory of Knowledge in Cognitive Perspective*. Pittsburgh, PA: University of Pittsburgh Press.

Rescher, N. 2003. *Epistemology. An Introduction to the Theory of Knowledge*. Albany, NY: State University of New York Press.

Rescher, N. 2006a. "Some Issues Regarding the Completeness of Science and the Limits of Scientific Knowledge," in Rescher, N. *Studies in the Philosophy of Science*, (pp. 57–81). Hausenstamm: Ontos Verlag.

Rescher, N. 2009. *Unknowability. An Inquiry into the Limits of Knowledge*. Lanham, MD: Lexington Books,

Rescher, N. 2012. "The Problem of Future Knowledge," *Mind and Society*, 11 (2), 149–163.

van Fraassen, B.C. 2008. "Rescher on Explanation and Prediction," in Almeder, R. ed. *Rescher Studies. A Collection of Essays on the Philosophical Work of Nicholas Rescher*, (pp. 339–361). Heusenstamm: Ontos Verlag.

The Limits of Information Science

Antonio Bereijo

Abstract

The problem of information science can be analyzed following several steps. First, there is the issue of the angle of analysis, which can be oriented in two directions: (a) to consider information science from the point of view of its boundaries as scientific endeavor (i.e., as a scientific rather than a non-scientific activity); and (b) to focus information science from the perspective of its possible confines in the present context as well as in the future.

Second, the limits of information science also depend on its own scientific status. In this regard, this paper characterizes information science as an applied science of design, i.e., as a kind of science of the artificial oriented toward the solution of concrete issues in the sphere of archives, libraries, and centers of documentation. This characterization as applied science of design involves the analysis of limits of the structural and dynamic elements of this science of the artificial, such as aims, processes, and results.

Third, besides the *endogenous aspects* of this problem-solving task in an artificial domain (which are commonly the case of aims, processes, and results), the limits of information science also depend on *exogenous features*, which are those elements (social, cultural, economic, political, ecological, etc.) that surround this scientific endeavor.

Keywords

Limits – information – science – endogenous – exogenous – artificial

Several aspects related to the limits of information science are analyzed here.[1] First, the angles of the problem of the limits of this discipline are specified. This includes taking into account the different ways to address this problem. Second, the configuration of information science as an applied science of design is studied in depth. This point of view leads to considering the limits of information science with regard to its aims, processes, and results. Third, some reflections are made on the general determining factors

1 Among the previous approaches to information science of interest are in Borko (1968); Cronin (2012); Griffith (1980); and Spink (2000).

 | DOI 10.1163/9789004325401_009

that limit information science: the endogenous elements and the exogenous components.

1 Angles of the Problem of the Limits

The problem of the limits of the information science can be analyzed from different angles. Two of these are highlighted here due to their general character: (i) the "borders" or boundaries that shape it as a scientific endeavor, and (ii) the "ceiling" or confines that it can reach as a scientific discipline oriented toward the solutions to specific problems. The reasons for this approach are clear. (1) We need to shape the boundaries or barriers that allow us to distinguish information science from the mere accumulated techniques or the plain accretion of professional practices (i.e., the knowledge based on personal experience or the knowledge that is shared by a professional group). (2) We have to think of what ceiling might be reached by the tasks of information processing and document retrieval (i.e., identification and re-identification of documents) in archives, libraries, and centers of documentation.

Regarding the first aspect – the boundaries or "barriers" that shape its scientific character – information science should be considered from the viewpoint of its being a dual discipline. This is because it is a discipline within the field of the artificial – it expands the human potentialities to access and recover information – [2] and, at the same time, information science is a discipline developed in a social environment, where the users constantly demand new and increasingly more sophisticated features and options (*prestaciones*).[3] Information science can also be characterized as an applied science, because it is a kind of science of design, insofar as it belongs to the artificial realm, which is oriented toward aims, processes, and results, as it searches for solutions to concrete problems.

Concurrently, there is a second angle regarding the final limit or "ceiling" that could be reached. Here it should be noted that there is complexity in information science,[4] due not only to the dual character – artificial and social – pointed out before, but also to the interrelation with other disciplines. The most obvious case is computer science, because of the necessity of automatized

---

2 "Sciences of the artificial" is understood here in the sense used by Herbert Simon. The clearest expression of this notion is in Simon (1996). On the role of rationality, see Bereijo (2003b).

3 On information science as a "dual science," see Bereijo (2011). On the features of the sciences of design as sciences of complexity, see Gonzalez (2013a).

4 Regarding the characterization of "complexity," see Gell-Mann (1995); Gershenson and Heyliguen (2004); Heylighen (1991 and 1999); Israel (2005); and Lyons (2005). On the existence of complexity in information science, see Bawden (2007); Broughton, Hansson, Hjørland, and López-Huertas (2005); Mukherjee (2008); and Vakkari (1999).

processing information systems. But complexity can also be found in the nexus of information science with other thematic fields such as computer science, linguistics, philosophy, history, etc. Thus, its scientific ceiling depends not only on the "internal" or specific capacity of the information science, but it can also be found in the dependency network that is established with other disciplines within a framework of inderdisciplinarity (not of multidisciplinarity).[5]

These angles of the problem of the limits are seen as more relevant insofar as it is acknowledged that this topic is an important issue in many facets. Some of these are scientific, such as the depth that can be reached by the research made in this thematic field or the confines of the pursued aims, the processes displayed, or the results that are finally achieved. Meanwhile, there are other facets, which are academic – such as the configuration of the curricula – or are related to the professional activity of the agents in the archives, libraries, and centers of documentation, in addition to being *de facto* the tools that make it possible to develop the features and options (*prestaciones*) intended for the users.

2 The Problem of the Limits of Information Science as an Applied Science of Design

When information science is analyzed as an applied science of design, the angles that have been pointed out – the barriers and the confines – are intertwined. On the one hand, in this case it is highlighted that there is a process of "scientification"[6] by means of which a series of professional practices have been led to designs of a scientific character. This has led to models oriented toward an articulation of aims, processes, and results. And, on the other, the epistemological and methodological ceiling of information science is determined precisely by its ability to elaborate models that are capable of solving more and more sophisticated problems.[7] These models of the applied designs are demanded by users with aspirations which are progressively expressed as maximizing (at least regarding the future).

Both from the angle of the barriers and from the perspective of the confines, we should think that there are three levels of rationality in this science

5 Concerning these issues, see Holland (2008); Moran (2006); Repko (2005); Rogers, Scaife, and Rizzo (2003); and Youngblood (2007).

6 On the concept of "scientification" and the features of the applied sciences, see Niiniluoto (1993).

7 In this regard, there is a resemblance to applied economics, as it is a science of design within the framework of the sciences the artificial, cf. Gonzalez (2008).

of design that should be taken into account:[8] (a) the general level of science, insofar as there is a scientific rationality that can be shared by the different sciences; (b) the specific level of the information science, as a rationality oriented toward the solution to concrete problems that are considered in archives, libraries, and centers of documentation; and (c) the rationality of the agents, which should make decisions (and, consequently, they are required to use the knowledge, practices, and values that are typical of scientific activities, in general, and documental activities, in particular).

Currently, information science is configured as a set of disciplines that share a common ground: they are oriented toward the solution to practical problems regarding documents. They are often problems related to knowledge organization. Consequently, information science deals with the organization of the documented knowledge: it usually deals with things that are registered by means of some representation system in some kind of backing or support (*soporte*).[9] Thus, when we try to establish the limits of information science as an applied science of design, the attempt to establish the limits as barriers or borders requires distinguishing the scientific components as opposed to the elements that do not have the features of a science. Thereafter, this affects the terminal limits or frontiers that shape this disciplinary field, which is certainly of a complex nature.[10]

When information science is characterized as an applied science of design, a dual approach is available: (a) the study made from a historical point of view; and (b) the analysis according to a thematic perspective (Bereijo 2011). Regarding the first approach, information science is oriented from its very beginning to the achievement of aims that are usually of a practical character. This claim can be reinforced nowadays by the fact that information science is connected with specific kinds of technologies. Thus, the limits of information science should be sought according to the three aforementioned parameters: (i) its aims; (ii) the processes that it displays; and (iii) the results that are finally achieved.

2.1 *Limits Regarding Aims*

Regarding *aims* of information science, it should be highlighted that these are usually related to the solution to practical problems that are connected to informative processes. Strictly speaking, information science does not seek

8 On these levels at issue in the sciences of design, see Gonzalez (2007).

9 This kind of knowledge is related to a documental realm that goes far beyond the physical base of the documents analogical and conventional, cf. Lyman and Varian (2003).

10 The complex nature of information science is the starting point of the analysis made by A. Bereijo (2012).

knowledge that is characteristic of basic science, like the knowledge directed to explaining or predicting more things and better. Furthermore, *historicity* is a relevant feature,[11] which is ever-present in knowledge organization. This feature can affect information science as a limit, insofar as the aims are always influenced by a temporal component of historical character. Besides this limitation, there is another factor that also involves limits to information science towards the future: its being intertwined with other disciplines. Because information science frequently uses elements that it does not produce, so it has a relevant dependence on other disciplinary fields, such as computer science, mathematics, linguistics, sociology, etc.[12]

As usually happens with regard to the technologies – on which information science is intensely dependent, relying above all on ICT – the *aims* of information science are commonly contextual, which means that they are "fruits of the moment" (*fruto de su tiempo*). The aims vary "according to the moment" of the research, which involves historicity in the approach and development of information science. These changes arise due to the need to solve specific problems in a given context. Consequently, the aims are constantly open to changes. These changes are due to internal reasons, as well as to the dependence on the external environment. In fact, information science continuously uses technological supports (*soportes tecnológicos*) that always modulate the possibility of achieving the aims that are sought. *De facto*, information science always has a set of related technologies that usually have an instrumental character for its advancement.

To be sure, some of the limits related to the *aims* of information science are *internal* and others are *external*. With regard to the former, they are usually related to the information that is recorded in some documentary support (*soporte documental*) or support. Normally, they deal with some of the following aspects: (a) the processes of selection of the documented information; (b) the ways and mechanisms to access the recorded information; (c) the management of informative records and their later *retrieval*; (d) the forms (*formulas*) and mechanisms oriented toward the distribution of the information for the human agents that act as information consumers; and (e) the study of the mechanisms of scientific communication (bibliometrics-scientometrics) and the analysis of the profiles of the users.

11 On the role of historicity in science, see Gonzalez (2011).

12 But this situation does not appear in the other direction, i.e., information science shows little ability to incorporate its own scientific research in other realms of knowledge. Tefko Saracevic claims that the advancements in information science have a very limited impact on other fields. Cf. Saracevic (1997, pp. 16–27; especially, p. 25).

At the same time, there are a series of limits that have their origins in *external factors*, where there is not only an increasingly demanding dependence on the users and institutions, but also a dependence due to the need to use technologies. Because of the contributions of the technologies, there is an increasing diversity of documental formats. In effect, in recent decades – from the 1990s on – there have been relevant changes. These are frequently related to the fast development of the information and communication technologies (ICT), especially through the Internet and the World Wide Web.[13] But they are advancements – with their respective limits – that originate outside the genuine documental field.

As information science seeks goals of a practical character, processes are required according to some criteria. They should be based on theories capable of solving concrete problems in order to achieve these practical aims. In this regard, Paul Thagard maintains that a theory is not scientific in the following cases: (1) when the theory is less progressive than its alternative theories over a long period of time, and so has not-solved problems; and (2) when the community of practitioners makes little attempt to develop the theory in order to solve problems, so that the community does not attempt to evaluate the theory in relation to others and is selective in considering confirmations and disconfirmations (Thagard 1978, p. 228). This double distinction seems to point to a *demarcator* that is related to information science. Because, when it is conceived as an applied science of design, information science has developed theories oriented toward solving practical problems.

2.2 *Limits Due to the Processes*

Certainly, the limits due to the processes are connected to the limits related to the aims. In addition, it happens that, on the one hand, the processes in information science depend on other scientific disciplines, among which computer sciences should be emphasized; and, on the other, these processes clearly depend on technologies based on the Internet, in general, and the World Wide Web, in particular. The former – the interdisciplinary relations – contributes to increasing the complexity in the realm of information science. In this sense, Wand has pointed out that the relations between information science and computer sciences have become more complex now than ever (Wang 2009). This can be seen, for instance, in the tasks of knowledge management, data-mining or reuse of information (*reutilización de información*). With regard to the relations between information science and

13 On the grow dynamics of the World-Wide Web, see Huberman and Adamic (1999). On the role of complexity in it, see Tredinnick (2009).

information and communication technologies (ICT), it can be claimed that these relations are in terms of possibility conditions for many processes of the information science.

Here, when we deal with the processes, the scientific character of the information science is also settled. When information science is developed, it offers solutions to concrete problems. These solutions are offered after evaluating theories and testing research processes, which have led to more efficient and effective ways of achieving the proposed goals. Nevertheless, there are spheres of knowledge within information science that can be considered scientific; meanwhile there are others that until now are just "technique" based on an accumulated professional *praxis*.[14] Disciplinary fields as knowledge organization or bibliometrics have already acquired a scientific status. This is so because an intense task of elaboration of theories oriented toward prescriptions to solve practical problems has been done, leading tools oriented toward the elaboration of specific paths to conduct the professional activity.

In effect, as information science is an "applied" discipline, it must combine prediction and prescription – with their respective processes to know the possible future and to give patterns to solve the problems considered – in order to have a scientific character.[15] Thus, the limits of information science derive from its ability to develop its own epistemological status as an applied science of design. Its limits are related to two kinds of conditions: (i) the internal conditions, that are usually linked to the advancement of knowledge in order to achieve the proposed aims (where predictions and prescriptions intervene); and (ii) the external factors, that are related to satisfy the informative demands of the users. These new demands are often associated with changes in the environment, which are usually stimulated by new technological advancements with regard to information storage and distribution.

But it is also necessary to demarcate what is "applied science" and what is just "application of science" (Gonzalez 2013b). Thus, in the case of knowledge organization, the task of designing and developing thesaurus or alphanumeric systems of classification is not the same as their practical application. This distinction can also be seen in the case of formal cataloguing tasks: the activity of elaborating cataloging standards (*estándares catalográficos*) is different from its practical application. The same happens with regard to the activities

14 It is just a professional praxis that occasionally rest on tradition, without theoretical criteria that articulate prediction and prescription. On the role of prediction and prescription in information science, see Bereijo (2003a, 2007a, and 2007b).

15 With regard to this feature, there is a resemblance to economics as applied science. On this issue, see Gonzalez (1998). Regarding these disciplines, cf. Bereijo (2011).

carried out in the realm of bibliometrics, when they are considered as a set of methods to analyze quantitatively the scientific and technological literature or, in the words of Alan Pritchar, as "the application of mathematics and statistical methods to books and other media of communication" (Pritchard 1969, p. 349). This is different from the activity of those that use that specific knowledge. In this regard, the mere application of statistics (*beam-counters*) does not make a field of knowledge scientifically rigorous (*científicamente riguroso*).

From a historical point of view, it seems that the origins of information science came about as a response to the inefficacy of library science with regard to the effective documentary treatment of an especial typology of documents, which is made for very concrete profiles of users: those users devoted to scientific research. In this way, library science stopped being useful to the documentary processing of this kind of specialized literature. Nowadays, it can be pointed out that we are witnessing a repetition of the phenomenon, since it is possible that the new technological developments result in process of disintermediation (*desintermediación*) [Nicholas 2012], so that they could supersede the current professional profiles in this field of knowledge.[16]

2.3 *Limits Regarding the Results*

The limits regarding the results or "products" of information science depend on the viability of the suggested processes and the proposed aims. Here, once again, the limits of science, in general, and the limits of the agents that make information science – whose rationality is usually bounded[17] – intervene. In this regard, it should be emphasized that this discipline displays a great variety of processes to achieve the proposed aims. This can be clearly seen in one of the more relevant fields of information science: knowledge organization. This is because, in knowledge organization, documental processes are displayed that are often modulated by means of standards to achieve relevant results, such as the organization, retrieval, access, and distribution of the documented information.

16 There are several reasons for this: (a) the impact of technologies that are usually related to the World Wide Web and telecommunications, that evolve rapidly to the satisfaction of the final users (Front-End/User-Friendly Technologies) in terms of "accessibility" or "usability"; (b) the increase of the information volume in the web; (c) the number of documentary typologies of an electronic character; and (d) the ubiquity of informative resources.

17 On bounded rationality in the context of social sciences, see Simon (2000). On scientific rationality in general, see Rescher (1999).

These kinds of processes are subject to limits of diverse character. They deal with the following elements: (1) the wide heterogeneity and volume of primary documents, which are in exponential growth (i.e., their wide typological diversity, their increasing volume, etc.); (2) the human agents that make documental processes, since they use procedures that are many times of a high intellectual complex nature; (3) the tools used, which are usually of a technological base (such as data bases, codification systems, etc.), but where there are also elements of prescriptive character that have been elaborated to guarantee the achievement of documentary homogeneous products; and (4) the limits imposed by the demands of the final consumers. Their requests are commonly subject to changes over the time, so that they vary greatly in accordance with many parameters. In many cases, their profiles are different and vary in accordance with cultural parameters. Usually they lead to the acquisition of new skills in the use of computer devices.

If the analysis is made in ontological terms, so that we see science as a human activity among others, oriented toward more and more sophisticated aims,[18] then the results can be seen in the following way. First, the result of the scientific activity in the field of information science is shown in solutions to concrete problems that are currently posed or that may be posed in the future. Second, as it is an applied science and is supported by design, it receives feedback form problems posed in the professional activity, with its wide typology of units of information. Third, because it is an empirical field, the proposed solutions to the problems posed must be tested in order to decide their validity. Thus, this research field should offer potentially universal or *universalizable* solutions,[19] instead of offering just *ad hoc* solutions or a mere solution of concrete cases without any further projection.

3 On the General Determining Factors that Limit Information Science: Endogenous Elements and Exogenous Components

There are two wide kinds of determining factors that limit information science: the *endogenous elements*, which are the elements that have been highlighted so far (the relation among aims, processes, and results), and the *exogenous*

18 On the components of a science, in general, and on science as human activity, in particular, see Gonzalez (2005, pp. 3–49; especially, pp. 9–11).

19 Nevertheless, it should be noted that there are limits to methodological universalism. See, in this regard, Gonzalez (2012).

components, which are those that surround the activity of this discipline, insofar as it is placed in a cultural and social environment that is historically changeable. Both kinds of factors also have a historical character: they evolve over time and, sometimes, there are radical changes (such as the revolutionary variations introduced in recent decades by means of ICT, in general, and the Internet, in particular).

With regard to the endogenous elements, the limits to computing the information should be highlighted, as well as the other limits due to its being a task made by human agents and the technological limits derived from the technological devices used by documentalists to information processing and transmission. Concerning the limits from the external viewpoint, the roles played by the users and the tasks of the institutions should be highlighted. Thus, the key is not the professional practice, but the limits of the activity that is supported by a scientific knowledge of an applied character, that is projected into the future through prediction and modulated by the patterns of prescription that is oriented to problem-solving.

It seems clear from a historical point of view that information science – as well as its precursor, library science – emerged in order to seek solutions to the problem of accessing and organizing the documented information. Thus, from their origins, these applied sciences have tried to offer answers to the problem of organizing and accessing the available documentary resources. They had to deal with their continuous growth in the face of two genuinely human *limits* that have an internal nature: on the one hand, the constraints that affect the ability to recall large amounts of information (memory); and, on the other, the constraints related to the ability to compute the information.[20]

But there is here an underlying problem. It is a problem of complexity in the process of *interaction* that appears between the human agent and the documentary *corpus* that he or she produces. In effect, this underlying problem, which lies in the limits to the human abilities, was the origin of a specific field of knowledge and was also the origin of some professional profiles that have their own features. This interaction between the agent and the document leads to problems that have remained over time and have been increasing noticeably in the last 30 years, due to the expansion of the Internet (in particular, the

20 It can be recalled, in this regard, the *Selva Selvaggia* metaphor, that was used by Ortega y Gasset to pointed out the relevance of the documentary organization tasks, or the design proposal of *Memex* (Memory Extender), by Vannevar Bush, as the technological precursor of the current information search and retrieval systems, that was made from the interconnection of concepts.

two decades of the Web) and its related technologies. In addition, the great variety of electronic formats and the new documentary typologies originated during these decades have also affected the problem of complexity.

To solve this complexity, besides the elaboration of conceptual elements – with their respective theoretical developments – different technologies have been updated in an attempt to solve the problem of organizing and accessing documentary collections. Now it is considered how to order, select, and guarantee public access to the documentation in an electronic support; i.e., the new ways of knowledge representation. These have been formulated by means of diverse systems of significance. In addition, these systems can have a full-fledged development and specific characteristics.

In this regard, as happens in other fields, technologies have been designed that intend to emulate the working of the human brain in some key aspects, such as the process of information storage, organization, and retrieval. Here, one of its *external limits* in the field of the information science can be seen clearly: the inclusion of knowledge that comes from different fields (such as computer science, cognitive sciences, mathematics, philosophy) in order to find solutions to the problems.

This explains why usually many of the technological devices that are used in information science have been elaborated to solve problems that have arisen in other fields of knowledge. This is the case, for example, of the first computer applications in the library realm, which have not been applied to genuine documental tasks (such as making bibliographic catalogues), but to the automation of library loan systems. They derived from commercial computer programs oriented toward the management of goods.

Similarly, the early attempts to automate the processes of descriptive cataloging were made from systems oriented toward the diagnosis of blood disorders (MICYN). This explains why information science is entitatively a set of different disciplines and knowledge that received the advancements of other fields of knowledge. Its limits not only depend on endogenous elements, but also on exogenous determining factors. In this regard, information science involves interdisciplinarity and this is, undoubtedly, a source of complexity, a factor that modulates its future.

References

Bawden, D. 2007. "Information as Self-organized Complexity: A Unifying Viewpoint," in *Information Research*, 12 (4). Available in: http://InformationR.net/ir/12-4/colis/colis/colis31.html (accessed on 9.9.2013).

Bereijo, A. 2003a. "Predicción y prescripción en las Ciencias de lo Artificial: Incidencia en las Ciencias de la Documentación," in Gonzalez, W.J. ed. *Racionalidad, historicidad y predicción en H.A. Simon*, (pp. 279–309). A Coruña: Netbiblo.

Bereijo, A. 2003b. "Racionalidad en las Ciencias de lo Artificial: El enfoque de la racionalidad limitada," in Gonzalez, W.J. ed. *Racionalidad, historicidad y predicción en H.A. Simon*, (pp. 131–146). A Coruña: Netbiblo.

Bereijo, A. 2007a. "Caracteres de la predicción en las Ciencias de la Documentación: De la *Information Science* al Análisis Documental," in Gonzalez, W.J., ed. *Las Ciencias de Diseño: Racionalidad limitada, predicción y prescripción*, (pp. 217–235). A Coruña: Netbiblo.

Bereijo, A. 2007b. "La articulación de la prescripción en las Ciencias de la Documentación," in Gonzalez, W.J. ed. *Las Ciencias de Diseño: Racionalidad limitada, predicción y prescripción*, (pp. 267–285). A Coruña: Netbiblo.

Bereijo, A. 2011. "The 'Category of Applied Science:' An Analysis of its Justification from 'Information Science' as Design Science," in Gonzalez, W.J. ed. *Scientific Realism and Democratic Society: the Philosophy of Philip Kitcher*, Poznan Studies in the Philosophy of Science and the Humanities, (pp. 327–350). Amsterdam: Rodopi.

Bereijo, A. 2012. "La sobriedad de factores en las Ciencias de la Documentación (*Information Science*) en cuanto Ciencias de la Complejidad: Proyección en el Análisis Documental," in Gonzalez, W.J. ed. *Las Ciencias de la Complejidad: Vertiente dinámica de las Ciencias de Diseño y sobriedad de factores*, (pp. 267–287). A Coruña: Netbiblo.

Borko, H. 1968. "Information Science: What is it?," in *American Documentation*, 19 (1), pp. 3–5.

Broughton, V., Hansson, J., Hjørland, B., and López-Huertas, M.J. 2005. "Knowledge Organization," in Kajberg, L. and Lørring, L. eds. *European Curriculum Reflections on Library and Information Science Education, Royal School of Library and information Science*, (pp.133–149).Copenhagen.

Cronin, B. 2012. "The Waxing and Waning of a Field: Reflections on Information Studies Education," in *Information Research*, 17 (3), pp. 1–9. Available in: http://informationr.net/ir/17-3/paper529.html (accessed on 10.11.2012).

Gell-Mann, M. 1995. "What's Complexity?," in *Complexity*, 1 (1) 16–19.

Gershenson, C. and Heyliguen, F. 2004. "How Can We Think the Complex?," in Richardson, K.A. ed. *Managing the Complex: Philosophy, Theory, and Practice*, (pp. 1–13). Naples, FL: Institute for the Study of Coherence and Emergence/Information Age Publishing.

Gonzalez, W.J. 1998. "Prediction and Prescription in Economics: A Philosophical and Methodological Approach," in *Theoria*, 13 (32), 321–345.

Gonzalez, W.J. 2005. "The Philosophical Approach to Science, Technology and Society," in Gonzalez, W.J. ed. *Science, Technology and Society: A Philosophical Perspective*, (pp. 3–49). A Coruña: Netbiblo.

Gonzalez, W.J. 2007. "Configuración de las Ciencias de Diseño como Ciencias de lo Artificial: Papel de la Inteligencia Artificial y de la racionalidad limitada," in Gonzalez, W.J. ed. *Las Ciencias de Diseño: Racionalidad limitada, predicción y prescripción*, (pp. 41–69). A Coruña: Netbiblo.

Gonzalez, W.J. 2008. "Rationality and Prediction in the Sciences of the Artificial: Economics as a Design Science," in Galavotti, M.C., Scazzieri, R., and Suppes, P. eds. *Reasoning, Rationality, and Probability*, (pp. 165–186). Stanford: CSLI Publications.

Gonzalez, W.J. 2011. "Conceptual Changes and Scientific Diversity: The Role of Historicity," in Gonzalez, W.J. ed. *Conceptual Revolutions: From Cognitive Science to Medicine*, (pp. 39–62). A Coruña: Netbiblo.

Gonzalez, W.J. 2012. "Methodological Universalism in Science and its Limits: Imperialism versus Complexity," in Brzechczyn, K. and Paprzycka, K. eds. *Thinking about Provincialism in Thinking*, (pp. 155–175). Poznan Studies in the Philosophy of the Sciences and the Humanities, vol. 100. Amsterdam/N. York: Rodopi.

Gonzalez, W.J. 2013a. "The Sciences of Design as Sciences of Complexity: The Dynamic Trait," in Andersen, H., Dieks, D., Gonzalez, W.J., Uebel, Th. and Wheeler, G. eds., *New Challenges to Philosophy of Science*, (pp. 293–305). Dordrecht: Springer.

Gonzalez, W.J. 2013b. "The Roles of Scientific Creativity and Technological Innovation in the Context of Complexity of Science," in Gonzalez, W.J. ed. *Creativity, Innovation, and Complexity in Science*, (pp. 11–40). A Coruña: Netbiblo.

Griffith, B.C. ed. 1980. *Key Papers in Information Science*. N. York: Knowledge Industry Publications.

Heylighen, F. 1991. "Coping with Complexity: Concepts and Principles for a Support System," in *Systemica*, 8, Part 1, 39–55.

Heylighen, F. 1999. "The Science of Self-Organization and Adaptativity," in *The Encyclopedia of Life Support Systems*, (pp. 1–26). Oxford: Eolss Publishers.

Holland, G.A. 2008. "Information Science an Interdisciplinary Effort?," in *Journal of Documentation*, 64 (1), 7–23.

Huberman, B.A. and Adamic, L.A. 1999. "Grow Dynamics of the World-Wide Web," in *Nature*, 401 (9), 143–150.

Israel, G. 2005. "The Science of Complexity: Epistemological Problems and Perspectives," in *Science in Context*, 18 (3), pp. 479–509.

Lyman, P. and Varian, H.R. 2003. "How much Information?" Available in: http://www.sims.berkeley.edu/how-much-info-2003 (accessed on 3.12.2012).

Lyons, M. 2005. "Knowledge and the Modelling of Complex Systems," in *Futures*, 37, 711–719.

Moran, J. 2006. *Interdisciplinarity*. London: Routledge.

Mukherjee, I. 2008. "Understanding Information System Failures from the Complexity Perspective," in *Journal of Social Sciences*, 4 (4) 308–319.

Nicholas, D. 2012. “Disintermediated, decoupled and down,” in *CILIP Update*, 29–31. Available in: http://ciber-research.eu/download/20120501-CILIP_Update_March_2012_pp29-31.pdf (accessed on 13.7.2012).

Niiniluoto, I. 1993. “The Aim and the Structure of Applied Research,” in *Erkenntnis*, 38, 1–21.

Pritchard, A. 1969. “Statistical Bibliography or Bibliometrics,” in *Journal of Documentation*, 25 (4), 349–349. Available in: http://independent.academia.edu/AlanPritchard/Papers/602982/Statistical_bibliography_or_bibliometric (accessed on 8.9.2012).

Repko, A. 2005. *Interdisciplinary Practice: a Student Guide to Research and Writing*. Boston: Pearson Custom.

Rescher, N. 1999. *Razón y valores en la Era científico-tecnológica*. Barcelona: Paidós.

Rogers, Y., Scaife, M., and Rizzo, A. 2003. “Interdisciplinarity: an Emergent or Engineered Proccess?,” *CSRP Technical Report*, v. 556. Brighton: University of Sussex.

Saracevic, T. 1997. “Users Lost: Reflections on the Past, Future, and Limits of Information Science,” *Newsletter ACM SIGIR Forum*, 31 (2), 16–27. DOI: 10.1145/270886.270889

Simon, H.A. 1996. *The Sciences of the Artificial*, 3rd ed. Cambridge, MA: The MIT Press.

Simon, H.A. 2000. “Bounded Rationality in Social Science: Today and Tomorrow,” in *Mind and Society*, 1 (1), 25–39. Spanish version “Racionalidad limitada en Ciencias Sociales: Hoy y mañana,” in Gonzalez, W.J., ed. *Racionalidad, historicidad y predicción en Herbert A. Simon*, (pp. 97–110). A Coruña: Netbiblo.

Spink, A. 2000. “Toward a Theorethical Framework for Information Science,” in *Special Issue on Information Science Research*, 3 (2), 56–84.

Thagard, P.R. 1978. “Why Astrology is a Pseudoscience,” in Asquit, P.D. and Hacking, I. eds. *Philosophy of Science Association*, v. 1, PSA, East Lansing, Available in: http//www.cavehill.uwi.esu/bnccde/PH29A/thagard.html (accessed on 3.1.2013).

Tredinnick, L. 2009. “Complexity Theory and the Web,” in *Journal of Documentation*, 65 (5), 797–816.

Vakkari, P. 1999. “Task Complexity, Problem Structure and Information Actions: integrating Studies on Information Seeking and Retrieval,” in *Information Processing and Management*, 35, 819–837.

Wang, B. 2009. “Information Science: The Territory and Relations to Computing Disciplines,” in *Journal o Library and Information Science*, 35 (1), 4–22.

Youngblood, D. 2007. “Interdisciplinary studies and the Bridging Disciplines: a Matter of Process,” in *Journal of Research Practice*, 3 (2), 1–9. Available in: http://jrp.icaap.org/ondex.php/jrp/article/view/104/101 (accessed on 18.9.2012).

PART 4

The Limits from Inside and the Limits from Outside

Rescher and Gadamer: Two Complementary Views of the Limits of Sciences

Alfredo Marcos

Abstract

The question concerning the limits of science leads us to think about the very metaphor of the limit. We found that there exist different kinds of limits and different possible actions in relation to them. In sum, there are limits that configure science and others that constrain it. The first ones must be respected, the latter overcome. We explore these limits following the suggestions of Rescher and Gadamer. These authors present complementary views. The first one approaches the question from inside the science, while the second one sees the limits of science from an exterior cultural point of view.

Keywords

Limits of science – Nicholas Rescher – Hans-Georg Gadamer

1 Introduction

The question regarding the limits of science requires us to think about the metaphor of limit itself (Section 2) and suggests that we then try to think about science from that metaphor (Section 3). In order to do this, I propose a dialogue with Nicholas Rescher (1928-) and Hans-Georg Gadamer (1900–2002), both of whom have written lucidly on the limits of science. Both were born in Germany, but Rescher's career has been in Pittsburgh and he may be considered an "Anglo-Saxon" philosopher of science, while Gadamer is a "continental", influenced by Heidegger and focused on hermeneutics. Their perspectives are complementary. The former contemplates the limits of science from within, from the point of view of someone primarily interested in science, while the latter looks at science from without, from a standpoint of a more general interest in civilization as a whole. I finish with a conclusive summary, recapitulating the main ideas acquired during my research (Section 4).

 | DOI 10.1163/9789004325401_010

2 Thinking about the Limit

The word *limit* comes from the Latin *limes-limitis*, meaning the boundary path between two plots of land, an origin offering semantic characteristics worth considering. The boundary shapes and constitutes the plot – without it there would be no real plot. A plot's boundary distinguishes it from other plots, separating it from them while joining it to them. As a path, it has some physical or geographical width; it is not just a geometric line. Its width allows us to think of it as having grey areas, or no man's lands, likely to be a place for collaboration or conflict. The boundary is not just something "in sight", there for our contemplation. It is also something "to hand", which invites us to walk, explore, to go beyond it... More than that, it is an entity arising from our action. As Antonio Machado's well-known line goes, "The road is made by walking" (*Se hace camino al andar*) [Machado 2001, p. 186].

In principle, a boundary is not an abstract line, but a concrete entity immersed in a context of action, relative to an agent. The agent contributes a space of possibilities, of possible actions, of attitudes, of aims and duties, of feelings and values that depend on a certain ontology. One can feel comfortable within certain boundaries or limits that do justice to the nature of things. Or, on the other hand, we can feel limits as constraints, something wrong, perhaps unjustly imposed. In this second case, our attitude leads us to step beyond the limits. A limit is seen either as something positive, valuable, that contributes to constituting an entity, or as something negative, unjustly constraining the entity.

As we can see, together with the concept of limit, there comes to us a universe of attitudes, of feelings and values linked to ontological presuppositions. For this reason, some limits are experienced by the agent as self-realization or perfection – "Become who you are", wrote Pindar (518–438 BC) – and others as a constraint or frustration. Later we shall see the importance of these considerations when we talk about the limits of science.

The limit was originally a spatial entity, but the word has undergone many metaphorical displacements, among others into the ambit of time. Thus the *Oxford English Dictionary*, in second place, lists the definition: "One of the fixed points between which the possible or permitted extent, amount, *duration*, range of action or variation of anything is confined." Like "end" or "bounds", the word "limit" soon began to play in the fourth dimension. It also underwent a similar displacement into the world of abstractions. There, paths lose their geographical width, their "usability", to become mere geometrical lines, not thoroughfares, or, in mathematical parlance, limits that are by definition unreachable. Finally, let us take note of the displacement of the concept of limit into the sphere of capabilities. Again, the above definition includes "... range of *action* or variation of anything is confined..."[1]

If our journey along the Latin *limes* has enriched us, we may expect something similar from the Greek *horion*, which normal dictionaries translate as "limit" or "border". Now the time metaphor is obvious: our *hours*, as we count them, are boundaries in time. The Hours (*Horai*) in Greek mythology were the goddesses who ordered and governed nature, who controlled the beneficial changing of the seasons. From this standpoint, they are precisely the boundaries cutting us off from chaos and confusion.

Not far from *horion* are the Greek words *horama* and *horasis* (sight, vision), and *horizo* (to limit), where we can glimpse our own word "horizon". The horizon is a fleeting visual boundary, impossible to reach, but which affects our actions as an objective. Our sight marks the direction in which we walk, that is, forwards – towards the front. Therefore the notion of horizon is not only visual but also "agential", in the same way as the notion of frontier, linked with front.

To cite only two significant and mutually contradictory examples, Vannevar Bush, then director of the Office of Scientific Research and Development, in July 1945 sent a report to the President of the United States with the significant title *Science, the Endless Frontier*, while in 1970 Bentley Glass, once president of the American Association for the Advancement of Science (AAAS), gave a speech to his association called *Science: Endless Horizons or Golden Age?*, which concluded with the statement that the horizons of science were no longer infinite.

After their passage through Latin, Greek words like *horion* have served to name other forms of limitation, such as banks and shores. When we look towards the limits of science, perhaps we should remember Newton's famous words: "...I seem to have been only like a boy playing on the sea-shore, and diverting myself in now and then finding a smoother pebble or a prettier shell than ordinary, whilst the great ocean of truth lay all undiscovered before me" (Brewster 1855, col. 2, p. 407).[2] Or St Augustine's famous story, from which this image could easily have been taken.

But before concentrating on science, let us explore once more the sources of our notion of limit, this time looking at the Greek word *peras*. This term, probably linked with "period" (time) and "perimeter" (space), calls to mind the concept of *apeiron* (the unlimited), surrounded in turn with connotations of

1 My italics.

2 The etymological works that I have consulted do not relate the English word "shore" with the Greek *horion*. Their evident similarity may be due to a mere coincidence – poetic justice, perhaps.

all kinds – negative for their indeterminateness and difficulty of understanding and positive for their fertility and potency.

We now see that to think in terms of the limits of science is not the same as wondering whether science has limits. Perhaps, in a certain meaning of the metaphor, it does and in another it does not. Furthermore, it will be necessary to posit the question of the value of limits. And the answer to this last question will inexorably beg new questions about our actions. We shall try to tackle all this in the dialogue with Rescher and Gadamer.

3 Thinking Science from the Metaphor of the Limit

3.1 *The Limits of Science: A Look from the Inside*

Nicholas Rescher dedicated a book to the question of the limits of science.[3] In it he maintains, in the first place, that science is not everything, that outside science there are forms of knowledge and praxis that are perfectly valid and rational. There are areas in which we have cognitive and practical interests and which are completely outside the province of science. The author speaks here of "the limited province of natural science" (Rescher 1999, Ch. 15). In his own words, "there is no question that natural science is subject to domain-external incapacities. We must recognize that various important evaluative and cognitive issues lie altogether outside the province of science as we know it" (Rescher 1999, p. 250). To think the opposite would be tantamount to subscribing the ideology called scientism.

As we see, Rescher thinks in terms of territories, with their limits or borders. Science takes up one of these domains, but beyond it there is still life. Limits of this type, which make up the profile of science against the background of the lifeworld, or *Lebenswelt*, could be called *constitutive limits*.[4] In my opinion, these are blurred limits, more "geographic" than "geometric", for there will always be cognitive contents and actions of doubtful assignation. The constitutive limit is open to passage, for there must be passage between science and life and vice-versa, and positive, for, rather than constraining, it configures science.

3 Rescher, N. (1984). A revised second edition appeared in 1999, which I will quote from now on.

4 From now on I shall allow myself to alter Rescher's terminology. He speaks of *disabilities, limits, incapacities* and *deficiencies* of science (1999, p. 3). In fact, all these terms refer to different types of limits. It would therefore be interesting for the term itself to indicate what kind of limit we are dealing with. The change in terminology that I have adopted seeks to carry out this function.

As for communication between science and the rest of the lifeworld, Rescher's position is only partially satisfactory. He admits that there must be outlets from science into life. Indeed, he states on several occasions that science's evaluation criterion can only be its practical usefulness. But he does not accept that there is traffic the other way, from the lifeworld into science. Science, he states in a curious *in crescendo*, is autonomous, self-sufficient, sovereign (Rescher 1999, pp. 249–250). But if the *limes* delimits as much as it joins, if it is going to be permeable and passable, it must be so in both directions. Science will have to accept requests, supports and also – why not – restrictions arising in other spheres.

Let us finally observe that science has to be viewed as an integral part of the lifeworld, science is inside it, and not in juxtaposition to it. Science is a part of human action. This integration and connection with the rest of the lifeworld will be dealt with below in the dialogue with Gadamer. But the step that Rescher allows us to take is a major one, as he recognizes the existence of the constitutive limits of science and the legitimacy of some areas of knowledge and practices situated beyond them.

Let us now examine the second type of limit. For this, we shall consider only problems concerning science, leaving aside those that do not fall within its sphere. Well, one might think that theoretical reasons exist to state that science will never offer a perfect solution to all the problems in its domain. These theoretical reasons would form a second type of limit, the *theoretical limits* of science. Outside them would remain those scientific problems that science, for theoretical reasons, will never be able to tackle, much less solve. Within them we should have the problems that science, at least in theory, would be able to tackle successfully. It would be the terrain of *science that is possible in theory*. Now, Rescher argues at length that such theoretical limits do not exist (Rescher Chs. 6 and 7). For Rescher, science that is possible in theory would simply be science.

Nevertheless, this thesis has been widely debated. Let us see an example to show the kind of objections that may be raised to it. Chaitin has shown, inspired by the works of Gödel and Turing, that the random character of a mathematical sequence cannot be proven, that it is undecidable. This has consequences for natural sciences, as was recently demonstrated by the Complutense University physicist Fernando Sols (2011). From Chaitin's demonstration, it could be inferred that the question of the presence or absence of purpose in nature is also undecidable, for we shall never know whether a sequence of natural phenomena occurs at random or is directed towards an end. What can we say then? That the problem of randomness and teleology has nothing to do with the natural sciences, or that it does belong to their domain but for purely theoretical reasons cannot be tackled? If we choose the first option, Rescher's position is upheld, but not if we choose the second.

In any event, what we are interested in here is not so much whether theoretical limits exist or not, but the very concept of the theoretical limit. These limits, if they exist, would have a somewhat more restrictive nature than the constitutive limits, and probably more "geometric" than "geographic" profiles. Rescher contributes to clearing this up by distinguishing between theoretical limits and *practical limits*. These are the third type of limits to science. Science does not reach many of the problems within its domain for reasons of a practical nature. For example, the capabilities of the Large Hadron Collider of the CERN mark a practical limit. It is the largest and most powerful particle accelerator in the world. If it were necessary for an experiment to go beyond its capabilities, then for the moment it quite simply could not be done.

Frequently these limits, as Rescher suggests, may be reduced to economic terms. But practical limits cannot always be translated into money. For example, there are historical moments when the mathematics that a part of natural science would have required was not available. For natural science, this means a practical limit that does not only depend on financial investment. Practical limits may also be linguistic, moral, social, political, ecological and of many other types. In some cases, these limits must be overcome, and in others respected.

In short, science that is possible in theory is limited by practical factors, which determine what is *science that is possible in practice*. We know that part of the science that is possible in theory will never be possible in practice. But we cannot know which part. According to Rescher, there is no way of determining which definite problems will remain outside scientific development. Problems that today lie beyond the practical limits may not tomorrow. We have before us a horizon-type limit: it is always there but it moves as we advance. This type of limit, in its movement, responds to what Rescher calls the Kantian principle of propagation of questions: "Science emerges as a project of self-transcendence. It embodies an inner drive that always presses beyond the capacity limits of the historical present" (Rescher 1999, p. 18). These are blurred, changing limits. They function as a challenge and a frontier. They invite us to transgression, but this never actually happens. Practical limits take up an intermediate position regarding their positive (constitutive) or negative (constrictive) nature. They are between constitutive and theoretical limits on the one hand, and fallibility limits on the other.

Fallibility limits include our personal inoperativeness, organizational and institutional faults, our lack of attention, work or honesty, errors which we inevitably commit given our human nature (all too human!). There are the limits that separate science that is possible in practice from *effective science*, which comprises effective achievements. These fallibility limits also appear as a horizon. They cannot all be overcome, although each of them may be individually. That is, science will always be fallible and incomplete, but none of its definite errors is dictated by fatality. The attempt to overcome limits of this kind is necessary, as they are limits of a purely negative nature: they are constraints on the development of science that generate deficiencies.

We have now seen constitutive limits, theoretical limits, practical limits and fallibility limits. The right moment has now come to remember Pindar's maxim: "Become who you are". If science *is* what its constitutive limits mark out, but *has* only *become* what the fallibility limits allow, the difference between them could be called *Pindar's gap*. To bridge this gap is the ultimate, indispensable and unattainable task of scientific undertaking.

All the limits that we have so far found have a common origin. They derive from science itself, that is, from its constitution, and from the subject producing it: mankind and his special circumstances, his lifeworld, his environment, or *Umwelt*. But science has an intentional character; it refers to something outside itself, to something outside even the lifeworld, and produces knowledge of that something. That something that science seeks to refer to is the world itself, or *Welt*. And this objective pole of science also imposes limitations on science. Let us call them *objective limits*. These cannot be broken and have a positive character. They cannot be considered as defects of science. That is, our science can be neither more precise nor more complex than nature itself, its depth and breadth cannot go beyond the dimensions of nature itself. If there is indeterminacy in nature, in our science there will be uncertainty. We shall never predict what nature itself has not determined.

Objective limits cannot be broken. Moreover, they cannot be reached, either. This is due to the other limitations that we have identified. For example, science is constitutively conceptual, and what is real, as Rescher states, can never be conceptually exhausted (Rescher 1999, p. 75). Or, to put in in Shakespeare's classic terms: "There are more things in heaven and earth, Horatio, than are dreamt of in your philosophy" (Shakespeare 1985, act I, scene V).

It is true that a creative and metaphorical use of language can shorten the distance between *physis* and *logos*, but can never bring them to the point of identity. We can only aspire to increasing the similarity between being and thinking.[5] The identity between our conceptual system and reality is one of those limits, unattainable and fleeting.

---

5 This topic is treated more in depth in Marcos (2012, Ch. 6).

Let us add a final remark concerning objective limits that has implications for the relationship between the natural sciences and the human sciences. The attempt to subsume in the natural sciences *all* knowledge about the human being may lead to an inconvenient overstepping of bounds. According to Rescher, "Inflating the claims of science to the point where it is held to have all the answers about the condition of man, the meaning of life, or the objects of social polity is a dangerous move [...] Such an inflated view of capacities invites skepticism and hostility in the wake of the disappointment of expectations that is its inevitable consequence" (Rescher 1999, p. 247). This disappointment stems from the attempt to overcome limits which may be of an objective character, for "Man is a member not just of the *natural* but of the specifically *human* order of things" (Rescher 1999, p. 248).[6]

3.2 *The Limits of Science: Looking from the Outside*

> "*My whole philosophy is nothing but* phronesis" (Gadamer 2003, p. 54).

Technoscience has constitutive limits, as we have seen.[7] And beyond them there exists intelligent life. Beyond technoscience we come across the remaining fields of the sphere of knowledge, such as art and morals. And the sphere of knowledge appears, in turn, against a background of the lifeworld, to which it doubtlessly belongs, and in which it has to coexist with other respectable human realities, as committed as technoscience may be to true knowledge and rational action. Cultural traditions, emotions, philosophy, religion, politics, education, communication and many areas of human life, ranging from daily experience and common sense to, for example, sport, make up the lifeworld, just as technoscience does.

According to scientism, the limits of science coincide with those of rationality. As Gadamer sums up, "one sees rationality in the context of science and confined within its limits" (Gadamer 1979, pp. 8–9). This is what the neopositivists of the Vienna Circle, in their manifesto, called "The scientific conception

6 For a more extensive treatment of this point, see Marcos (2010).

7 Science and technology are distinguishable realities both historically and conceptually. Nevertheless, today, their level of symbiosis is such that we can speak correctly of technoscience. For a philosopher of science like Rescher, the conceptual difference between science and technology is very important, which is why we kept to it thus far. On the other hand, for Gadamer, whose interest lies more in hermeneutics than in the philosophy of science, the relevant entity is rather the amalgam of science and technology, which justifies our talking henceforth of technoscience. See Gadamer (1985–1999, vol. 4, p. 247 and 1996, p. 6).

of the world" (*Wissenschftliche Weltauffassung*) [Carnap, Hahn, and Neurath 1929].

But if what we seek is thought more in line with truth and in line with the relationships between technoscience and the other areas of human life, we must look for it a long way from scientism, in some philosophical tradition that shows more respect for those areas other than technoscience.

We have to overcome the modern idea of an absolutely autonomous science. We must reintegrate the sphere of knowledge, and technoscience in particular, into the lifeworld. Technoscience must interact with its environment. Therefore, it needs a healthy environment, made up of entities worthy of respect. It is only one facet of our life, which borders many others. In other words, one of the functional limits of technoscience is that it is not sufficient by itself to give a basis for an entire civilization, to a whole way of life.

To a great extent, this is the message of Gadamer's philosophy. This limit implies no deficiency of technoscience; it is not at all negative, except for those who, with a scientist mind-set, would base everything on technoscience. Gadamer's philosophy is not antiscientific. It is, however, anti-scientist. This is the main reason for choosing Gadamer as an interlocutor in the present text. But there are more reasons. Gadamer's arguments are very near to those of many other contemporary philosophers, whose echoes we shall hear together with Gadamer's voice. I mean other contemporary thinkers, in both the Anglo-Saxon and continental traditions and especially Heidegger, Arendt, Husserl, Dewey, Wittgenstein, Popper, Kuhn, Polanyi, Toulmin, MacIntyre, Putnam, Habermas and Ricoeur. A dialogue with Gadamer would be then, in some ways, like a dialogue with many of these writers' ideas. Like most of them, Gadamer identifies the limits of technoscience, points out it insufficiency as a single basis for a civilization, denounces the excesses of the scientism, and all this he does without defecting to the ranks of the antiscientific mind-set, without falling guilty of relativism or irrationalism, without going into what he himself calls "the shadow of nihilism" (Misgeld and Nicholson 1992, p. 114 and Gadamer 1985/1999, vol. 9, p. 367).

Gadamer's hermeneutics may be read as a theory of the limits of science, as Stefano Marino states (Marino 2011, p. 33, n. 37). Science does not use up all the territory of truth, of knowledge or of experience, nor everything can be achieved by its means (Gadamer 1993, pp. 127–128). His thought supposes a critique of the scientist *hybris* that would take science beyond its constitutive limits. As a complement of this *pars destruens*, there appears in his work a *pars construens*, which seeks the reassessment of other areas of "human experience of the world in general", which, according to Gadamer, "goes beyond the limits of the concept of method as set by modern science" (Gadamer 2004, p. xx).

"One cannot ignore such 'knowledge', in whatever form it expresses itself: in religious or proverbial wisdom, in works of art or philosophical thought" (Gadamer 2004, pp. 565–566). One has "to understand the variety of experiences – whether of aesthetic, historical, religious or political consciousness" (Gadamer 2004, pp. 84–85). These experiences are beyond the limits of science, and science must not attempt to colonize them. They must be respected and pondered for themselves, for they are by nature irreducible to the methods of science.

We may wonder now what concept of science this drawing of limits that Gadamer proposes depends on. Well, to characterize modern science, Gadamer turns to a few concepts with their roots clearly in Descartes and Bacon. The first of them is the concept of method. Modern science is primordially method. It is a method with a vocation for universality, for automatism and certainty. This notion takes on so much importance in Gadamer that it forms part of the title of his flagship book, *Truth and Method*. Around the 17th century, a new form of civilization arose, a new *way* or *form of life* (*Lebendsform*), defined almost univocally by the emergence of a new notion of science (Gadamer 1983, p. 6). The essence of that notion is summed up in a single word: method. The effect of method is objectivation, that is, the configuration or delimitation of the object, the transformation of (part of) reality into an object.

The method is, then, an objectifying one. It objectifies by delimitation. Therefore, science does not only have limits, but, at a deeper level, it *is* a limit, it is born of a process of delimitation. Our aim in the first place is to draw a limit between subject and object, along the lines of Descartes' separation of *res cogitans* and *res extensa*. We thus cut reality into two parts, which we set up against each other. One of them is the object for the other, and also inevitably an obstacle. It is what is not the subject, and what resists the subject. Immediately there arises from this arrangement of things the attitude of control, of dominion and planning as procedures for reintroducing the subject in objective reality. It is in this that the new way of life of the modern subject consists. The subject that has been separated from the object returns to it as the dominator. "'*Objekt*' or '*Gegenstand*' is defined through a 'method' that prescribes how reality is to be turned into an object. The aim of methodologically researching the object in this way is then essentially to break down the resistance of 'objects' and to dominate the processes of it" (Gadamer 1998, p. 127). Thus become imbricated science, which supposedly knows with Cartesian objectivity, and technology, which will contribute the Baconian control of the object. Thus is opened up the path to today's technoscience.

The possibility of technoscience overstepping, or *hybris*, is now clearly intuited. This will happen when we seek to impose the objectifying method and the attitude of dominion on all reality. We overstep our limits, we take technoscience beyond its constitutive limits, when we accept, in Gadamer's words, that "nothing can be scientifically investigated or truly understood, unless it conforms to procedure of method. Henceforth, objectivity in this sense specifies the very limits of our knowledge – what we cannot objectify we also cannot know" (Gadamer 1987, p. 41). This movement can be seen indifferently as an unjustified extension of technoscience or as an unjustified reduction of reality. Scientism and reductionism go hand in hand. However we look at it, the result is the same: the identification of the limits of human knowledge with those of the scientific method, and the consequent attempt to base all our actions, all our relations with reality, on the application of scientific knowledge.

We know the consequences. In the epistemic, objective truth is replaced by subjective certainty (Gadamer 1996, p. 148). In the practical realm, this is an attempt at the artificializing of all things natural. What starts as an objectifying movement turns into an immense subjectivizing of reality. Thus are laid the foundations for the uneasiness of our culture. But before entering the chapter of unease of modernity in depth, allow me to be yet more precise concerning the idea of objectifying by limitation. And the fact is that the methodical splitting of the real into subject and object was concreted also in other splits. We separate primary qualities from secondary ones, we carefully delimit the quantitative from the qualitative. We leave apart, of course, all emotional evocation, any aesthetic quality. We observe "facts" away from values. The scientific method seems to require it. We leave in parentheses, away from reality, everything aesthetic, emotional, qualitative, axiological, never to come back to it, to negate it or simply exclude it. Or to try to reduce it forcibly to the parameters of the objectifying method.

Let us now go on to the question of our western civilization's uneasiness in its modern version, which manifests itself through multiple symptoms, which became especially visible during the last century. I shall name some of them, identified explicitly by Gadamer himself, although surely each reader will be able to add a few more. As the main symptom of the modern pathologies we could identify what Gadamer calls the *shadow of nihilism* (Gadamer 1985/1999, vol. 9, pp. 367–382; vol. 3, p. 407; Misgeld and Nicholson 1992, p. 114). Under this poetic formula we could include the atmosphere of anxiety that dominates modern life, together with the lack of hope and meaning of life that technoscience is unable to alleviate. We must also include the vacuum left by the dissolution of religion, effected by the scientist mentality, a vacuum that technoscience is incapable of filling (Gadamer 1993, p. 197 and 1996, p. 159). According to Gadamer, "the contribution of the scientific Enlightenment reaches an insuperable limit in the mystery of life and of death" (Gadamer

1996, p. 67 and 1985/1999, vol. 4, p. 293). In the same way, Gadamer identifies modern voluntarism and relativism as pathological symptoms (985/1999, vol. 10, pp. 236, 263), which lead to moral subjectivism (1985/1999, vol. 7, pp. 398–399) and aesthetic irrationalism (Gadamer 1986). Together with them we have fragmentarism and specialism (Gadamer 1985/1999, vol. 10, p. 263), individualism, lack of solidarity, the break-up of community sense (1985/1999, vol. 10, pp. 235–236 and 1989, p. 157) and others like consumerism (1985/1999, vol. 4, p. 256 and 1996, p. 18) or historicism (1985/1999, vol. 10, p. 263).

We have sought to base our way of life on technoscience, but this means clearly going beyond its constitutive limits. Technoscience does not go that far, it cannot sustain a way of life. As a result, the modern West has been beset with a number of ailments. If we want to cure our civilization, if we want our way of life – including technoscience – to survive in its postmodern version, we must find other bases for it. For this, Gadamer proposes the rehabilitation, together with technoscience, of other areas of knowledge, of experience and human action, and a dialogue between them all.

But more especially the German philosopher concentrates on the rehabilitation and autonomy of practical knowledge, *irreducible* to episteme or science. The role of the expert is always important, but the final decision in all our actions, even in those making up technoscientific research, corresponds rather to practical wisdom.[8] It is from this that our non-delegable responsibility derives. Practical wisdom is formed from the practices themselves. For example, to participate in a given tradition is essential for its formation: "We produce ourselves inasmuch as we understand [and] participate in the evolution of tradition" (Gadamer 2004, p. 293 and 1985/1999, vol. 1, p. 298).

In a way that might seem provocative to the modern mentality, Gadamer argues for the rehabilitation of the authority of tradition. To avoid any ambiguity, let us remember that tradition, for Gadamer, is an ongoing event, not a static entity. Therefore, there is no kind of defence here of the *status quo*. What is true is that continuity of a given long-lived tradition clearly benefits the rationality of technoscience. Scientific paradigms, as Thomas Kuhn has shown, arise in a seemingly discontinuous, revolutionary and somewhat disconnected way. One may question the commensurability and possibility of comparison between them, and therefore the possibility of justifying the rationality of scientific decisions and the very progress of science. However, the different scientific paradigms are, in fact, comparable in a rational way, as Kuhn himself maintains (Kuhn 1977, Ch. XIII), thanks to the persistence underlying the changes

8 This idea is examined in depth in Marcos (2010, Ch. 3).

of a communitary current of values, practices and wisdom that we might well call tradition and which goes far beyond the limits of science.

All this suggests an inversion: it is not only that technoscience is incapable alone of upholding our way of life but that, on the contrary, technoscience itself maintains its aspiration to rationality because it is based on certain practices, values and traditional kinds of wisdom characteristic of a certain way of life. There is a way of life that makes the appeal to reason possible. Gadamer proposes, in this regard, an inversion of the modern position, especially the Kantian position, which sought to base *ethos* on reason. In exchange, he recovers the Aristotelian perspective: it is a certain human *ethos* that allows for the development of rationality, including scientific rationality (Gadamer 1985/1999, vol. 4, pp. 187–188 and 1999, pp. 29, 34–35). "The rationality of practical reason receives its normative power not so much from arguments as from what Aristotle called '*ethos*', that means from the determination of one's emotional life that shows practical wisdom at work in education and moral training" (Gadamer 2000, pp. 48–49 and 1985/1999, vol. 8, p. 437).

The practical wisdom that Gadamer refers to is therefore situated in the Aristotelian tradition, and it would not be at all unfair to identify it with the intellectual virtue of *phronesis*. Indeed, in clear agreement with the *Nicomachean Ethics* (1106b 36 ff., 1144a 35–36), Gadamer states that "there is no *phronesis* without *ethos* and no *ethos* without *phronesis*" (Gadamer 1999, p. 155 and 1985/1999, vol. 7, p. 390. See Marcos 2012, Ch. 2).[9]

4 Concluding Summary

We have tried to think about science from the metaphor of the limit. For this, in the first place, we had to delve into the metaphor itself. We have seen the different levels of its metaphoricalness, from the most conventional, recalling boundaries in space and time, to the most metaphoric, referring to functional limits. The connotations inherent in the different versions of the metaphor are in turn very diverse. The limit is in some ways a positive concept, for it constitutes entities, while also being negative, insofar as it constrains them; it may be clear or blurred, static or dynamic, permeable or impermeable.

The metaphor of the limit is very useful for thinking about science. But we have seen that in fact it is insufficient for this task by itself. It is a metaphor that

9 As Gadamer suggests, if we had to look for a Kantian correlate for this type of practical wisdom we should have to look back to the *Critique of Judgement* rather than either of the other two great critical works. Cf. Gadamer (1985/1999, vol. 10, p. 278).

contributes clarity especially when inscribed in a web of metaphors. Ideas like border, exploration, path, bank, hour, horizon, link, nexus or pore are near to the idea of limit, they belong to the same metaphor web. But we have found all the metaphors of an *agential* nature especially interesting. The idea of limit takes us immediately to the idea of a subject that does things with that limit: it respects it, it travels along it, it reaches it, it crosses it, it chases it, it constructs it and explores beyond it.

Once we have gone deeper into the metaphor of limit and into its connections with nearby ones, we are then able to apply it to the task of thinking science. For this, we have used a dialogue with Rescher and Gadamer, as complementary authors.

With Rescher, we have identified the limits of science looking at it from the inside. Thus we have distinguished, in this order, constitutive, theoretical, practical and fallibility limits. Within the constitutive limits lies science as a reality and a possibility, that is, all science. Within the theoretical limits we would find science that is possible in theory. Within the practical limits we have science that is possible in practice and within the fallibility limits, effective science. Scientific undertaking as a whole is guided by the aspiration to reduce the difference between effective science and just plain science. It is a question of bridging what we have called *Pindar's gap*, so that science *becomes* what it *is*. It is an indispensable and impossible task, tragic though that may sound. Science, moreover, also has objective limits, marked by the very nature of things and especially by human nature.

The constitutive limits of science, as Rescher sets them out, leave room for other equally respectable human realities. In other words, science is a part of the lifeworld, which is much larger than science itself. We are interested, therefore, in thinking about science's relationships with its surroundings, with other aspects of human life. We are interested in looking from the outside to the limits of technoscience. We have journeyed this stretch of the path in a dialogue with Gadamer.

According to him, technoscience is not enough to base a civilization on, to sustain a way of life. That is one of its limitations. However, western civilization, in its modern version, tried to seek a basis mainly in technoscience, on both the epistemic and practical planes. From the failure of that attempt there derives the unease of the modern West. That unease displays itself in a multitude of symptoms that we all know and which Gadamer poetically summed up in the expression "shadow of nihilism". We have examined some of those symptoms and have addressed the question of how to cure them, or at least how to alleviate them.

The most promising of the strategies consists in the rehabilitation of other areas of knowledge, of human action and experience and more especially the reassessment of practical wisdom, or *phronesis*. This is wisdom that gives support to an *ethos*, to a way of life, where value is given to appealing to reason and experience, and in which there is room for technoscience. But it should be remembered that this practical wisdom is based in turn on the *ethos* that the wisdom itself contributed to found. Nobody should be shocked by a (hermeneutic?) circle of this sort, a reciprocal support between practical wisdom and sensible practice. Technoscience, for its part, far from founding a way of life, which is beyond its limits, is given a basis by it through practical wisdom.

References

Brewster, D. 1855. *Memoirs of the Life, Writings, and Discoveries of Sir Isaac Newton.* Edinburgh: Thomas Constable and Co.

Carnap, R., Hahn, H. and Neurath, O. 1929.*Wissenschaftliche Weltauffassung: der Wiener Kreis*. Vienna: Artur Wolf Verlag.

Gadamer, H.G. 1979. "Historical Transformation of Reason", in Geraets, T.F. ed. *Rationality Today. La rationalité aujourd'hui. Proceedeings of the International Symposium*, (pp. 3–14). Ottawa: University of Ottawa Press.

Gadamer, H.G. 1983. *Reason in the Age of Science*. Cambridge, MA and London: The MIT Press.

Gadamer, H.G. 1985/1999. *Gesammelte Werke*. Tübingen: Mohr Siebeck.

Gadamer, H.G. 1986. *The Relevance of the Beautiful and Other Essays*. Cambridge: Cambridge University Press.

Gadamer, H.G. 1987. "The Relevance of Greek Philosophy for Modern Thought," in *South African Journal of Philosophy*, 6 (2), 39–42.

Gadamer, H.G. 1989. *Das Erbe Europas. Beiträge*. Frankfurt a. M.: Suhrkamp.

Gadamer, H.G. 1993. *Über die Verborgenheit der Gesundheit. Aufsätze und Vorträge.* Frankfurt a. M.: Suhrkamp.

Gadamer, H.G. 1996. *The Enigma of Healing: The Art of Healing in a Scientific Age*. Stanford: Stanford University Press.

Gadamer, H.G. 1998. *Praise of Theory: Speeches and Essays*. New Haven and London: Yale University Press.

Gadamer, H.G. 1999. *Hermeneutics, Religion and Ethics*. New Haven and London: Yale University Press.

Gadamer, H.G. 2000. "Towards a Phenomenology of Ritual Language," in Schmidt L.K. ed. *Language and Linguisticality in Gadamer's Hermeneutics*, (pp. 19–50). Lanham and Oxford: Lexington Books.

Gadamer, H.G. 2003. *A Century of Philosophy. Hans-Georg Gadamer in Conversation with Riccardo Dottori*. London and New York: Continuum.

Gadamer, H.G. 2004.*Truth and Method*. London and New York: Continuum.

Kuhn, Th. 1977. *The Essential Tension. Selected Studies in Scientific Tradition and Change.* Chicago: The University of Chicago Press.

Machado, A. 2001. *Poesías completas*. Barcelona: RBA.

Marcos, A. 2010. "Filosofía de la naturaleza humana," *Eikasia. Revista de Filosofía*, 6 (35), 181–208. Available on www.revistadefilosofia.com, (accessed on 6.4.2016).

Marcos, A. 2010. *Ciencia y acción. Una Filosofía práctica de la Ciencia*, Colección Breviarios. México D.F.: Fondo de Cultura Económica.

Marcos, A. 2012. *Postmodern Aristotle*. Newcastle: Cambridge Scholars Publishing.

Marino, S. 2011. *Gadamer and the Limits of the Modern Techno-Scientific Civilization*. Bern: Peter Lang.

Misgeld, D. and Nicholson, G. eds. 1992. *Hans-Georg Gadamer on Education, Poetry and History: Applied Hermeneutics*. Albany: SUNY Press.

Rescher, N. 1999. *The Limits of Science*, (revised edition). Pittsburgh: University of Pittsburgh Press. (1st edition: Berkeley: University of California Press, 1984.)

Shakespeare, W. 1985. *Hamlet, Prince of Denmark*, edited by Ph. Edwards. Cambridge: Cambridge University Press.

Sols, F. 2011. "Heisenberg, Gödel y la cuestión de la finalidad en la Ciencia," contributed paper to the *Simposio Internacional Ciencia y Religión en el siglo XXI: ¿diálogo o confrontación?* Madrid: Fundación Ramón Areces.

The Obstacles to Scientific Prediction: An Analysis of the Limits of Predictability from the Ontology of Science[1]

Amanda Guillan

Abstract

The ontological obstacles to scientific prediction are a crucial issue regarding the limits of science. This issue is connected with complexity, both structural and dynamic. In this regard, the analysis of predictability from the ontology of science follows two steps. First, there is a characterization of the ontological obstacles to scientific prediction. Second, an analysis of these ontological obstacles is made, focusing from the perspective of complexity.

Concerning the first issue, the framework provided by Nicholas Rescher is relevant. It takes into account several ontological limits: (i) anarchy and volatility; (ii) chance, chaos, and arbitrary choice; (iii) creativity; and (iv) the limits on predictors. Regarding the second aspect, the ontological obstacles to predictability as a source of complexity for predictions should be emphasized. Certainly, scientific prediction can deal with a complex reality (natural, social, or artificial), so complexity can have influence on the possibility of prediction as well as on the type of prediction that could be attainable. Thus, the perspective of complexity – both structural and dynamic – is crucial when scientific prediction is analyzed from the ontology of science.

Keywords

Ontological obstacles – scientific prediction – complexity – historicity

1 A Characterization of the Ontological Obstacles to Scientific Prediction

A crucial issue regarding the limits of science is the problem of the obstacles to scientific prediction. Within the limits of prediction, the ontological obstacles (i.e., the limits due to the reality of the predicted phenomena) should be

1 This research is supported by the *Program* FPU of the Spanish Ministry of Education, Culture and Sport.

 DOI 10.1163/9789004325401_011

highlighted. Certainly, the knowledge of the extramental reality is an indispensable aspect of prediction, since predicting is a cognitive anticipation of a possible future. In effect, prediction "is a cognitive venture whose successful pursuit is inseparably bound up with factual matters regarding the nature of the world's *modus operandi*" (Rescher 1998a, p. 81). In this case, the ontology of science is in the service of the epistemology and the methodology of science. Obviously, it is an important field within the study of scientific prediction and its limits.

Due to this unbreakable nexus between the reality of the phenomena and the success in the prediction about those phenomena, it seems clear that the reliability and credibility that must characterize a successful prediction only can be achieved under certain conditions. In this regard, Nicholas Rescher has offered a quite interesting proposal on the ontological obstacles to predictability. Moreover, one of his main contributions to the research about scientific prediction is, in my judgment, the identification and analysis of the obstacles to predictability (above all, epistemological and ontological). Within the ontological obstacles, some of them might be highlighted: (i) anarchy and volatility; (ii) chance, chaos, and arbitrary choice; and (iii) creativity.

1.1 *Anarchy and Volatility*

There are ontological limits that affect prediction, mainly because prediction deals with phenomena that have not happened yet and, therefore, these phenomena are still open to the future. Rescher highlights this feature when he considers the obstacles to prediction. In his judgment, "ontological limits exist insofar as the future of the domain at issue is *developmentally open* – causally undetermined or underdetermined by the existing realities of the present and open to the development of wholly unprecedented patterns owing to the contingencies of choice, chance, and chaos" (Rescher 1998a, p. 134).

If the preconditions of the rational process of prediction are considered, it can be clearly seen that complexity of phenomena and events set limits to our predictive ability (Gonzalez 2010, pp. 271–274). In effect, from a methodological point of view, Rescher admits a plurality of processes that can lead us to achieve predictions. This methodological plurality responds, in the first place, to the kind of question that is posed; and, in the second place, it is due to the diversity of phenomena that we want to predict (Rescher 1998a, p. 245).

Independently of the methodological pattern that is used, Rescher demands that every predictive procedure meets several requirements. (1) *Data availability* (the relevant data about the past-and-present happenings can be obtained). (2) *Pattern discernability* (there are discernable patterns that can be detected in the body of data). (3) *Pattern stability* (since the prediction success largely depends on the stability of the past-and-present patterns, so these patterns continue into the future) [Rescher 1998a, p. 86].

Thus, for Rescher "rational prediction pivots on the existence of some sort of appropriate *linkage* that connects our predictive claims with the input data that provide for their justification" (Rescher 1998a, p. 87). In this way, he brings the ontological level to the forefront, as a support to the epistemology and methodology of science. In that case, the complexity of the reality conditions the complexity of knowledge and the consequent method in this regard. This is because we do not want just predict, but achieve reliable, accurate, and precise predictions.

This feature leads to admit the existence of ontological obstacles to predictability, since the very nature of phenomena can make it difficult or even impossible to meet the preconditions of rational prediction: the availability of the relevant data, pattern discernability, and the stability of the patterns followed by the phenomena. From this ontological perspective, a first distinction that we should take into account has to do with the separation between unpredictability and non-predictability.[2]

There is "unpredictability" when the nature of phenomena makes it impossible the prediction (now and in the future) so that it involves the *complete* impossibility to predict. Meanwhile, "non-predictability" has to do with the *current* impossibility to achieve a prediction, on the basis of the knowledge or the process of prediction that are available in a given moment. Thus, the former is a methodological-ontological issue (the very reality of the phenomena and events makes the prediction impossible); while the later involves a methodological-epistemological feature (the future knowledge might allow us to overcome the current difficulties) [Gonzalez 2015, pp. 54–56].

Unpredictability can appear as a result of the presence of phenomena of anarchic character (Rescher 1998a, pp. 136–138). Because, in this case, there are no regularities that allow the inference oriented towards the future, so the second precondition of the rational process of prediction cannot be met (i.e., pattern stability). However, it seems that, in principle, it can be too drastic to claim that there is a phenomenon or event clearly unpredictable. This difficulty is due to the dynamic character of scientific knowledge, which involves historicity. From this perspective, it is problematic to give examples of unpredictable issues without taking into account the historical dimension of knowledge (Gonzalez 2015, p. 56).

2 On this distinction, see Gonzalez (2010, p. 289).

According to Rescher, only science can inform us of its own limits,[3] so that it is not the society what sets the limits. This position involves that the limits of science are always with regard to a historical moment of scientific development. For this reason, even when he admits that there are anarchic phenomena, which can be intractable from a predictive viewpoint, he takes a wary attitude with regard to the existence of "unpredictability" in the strict sense. This is because, in his judgment, we can only claim that a phenomenon is anarchic on the basis of the available knowledge about that phenomenon, which can change in the future.[4]

In contrast to "unpredictability," "non-predictability" has to do with the *current* impossibility to obtain a prediction. In Rescher's account of the ontological obstacles to scientific prediction, "non-predictability" is related to the volatility of the phenomena or events (Rescher 1998a, p. 79). Usually, the current impossibility to predict is due to the instability of phenomena, so that the discernible patterns from the past can vary in the future (Rescher 1998a, pp. 79–82).

When volatility is considered as an ontological limit to predictability, the time dimension is crucial. Thus, frequently the processes are more stable in the short run than in the long run. For this reason, when prediction is in the long run, it is more likely that errors appear; while predictions in the short run are generally more reliable. Here the information used is another factor to consider, since the available information to predict in the short run is generally more relevant than when prediction is in the long run (Granger 1989, p. 7).

But it should be pointed out that volatility of phenomena or events is a matter of degree. In this regard, Rescher suggests a classification of the volatile processes in two groups: (a) moderately volatile processes (for example, the weather in the temperate regions); and (b) very volatile processes, which are characterized as processes which exhibit sudden and fortuitous changes (for example, wind velocity in the English Channel) [Rescher 1998a, p. 79]. Thus, a prediction in the long run that is about a process moderately volatile can be more reliable than a prediction in the short run about a process very volatile.

1.2 *Chance, Chaos, and Arbitrary Choice*

In Rescher's judgment, the main ontological obstacles to predictive success are chance, chaos, and arbitrary choice (Rescher 1998a, p. 134). He describes chaos and chance as stochastic or random processes that make the laws at issue probabilistic laws (Rescher 1998a, p. 134). At the same time, arbitrary

3 Cf. Rescher, N., *Personal Communication*, 17.6.2014.

4 Cf. Rescher, N., *Personal Communication*, 1.7.2014.

choices can be considered as another important obstacle to prediction. This is because, in Rescher's approach, predictability of human actions depends, to some extent, on their rational character (Rescher 1995, pp. 327–332). Thus, those actions and choices that are arbitrary (and consequently non-rational) can be, in principle, not-predictable.

Certainly, the presence of chance is an obstacle to predictability. This is because, "where chance is at work, the world can exhibit a fixed past and nevertheless confront us with distinct but altogether feasible futures – situations that are descriptively indistinguishable may unfold in different ways and issue in totally different results" (Rescher 1998a, p. 138). From this perspective, chance does not make it impossible to predict, but it affects the possibility to achieve predictions with certainty (Rescher 2009, p. 108).

From this point of view, chance does not involve unpredictability in a strict sense, but it may involve non-predictability. However, the degree in which a phenomenon or event characterized by chance is not-predictable is an issue that depends, in turn, on the kind of predictive question posed about that phenomenon or event. As Rescher points out, when there is chance, it is usually possible to achieve generic predictions and predictions on a large scale: "Individual chance events are indeed unpredictable, but the very randomness of chance fluctuation means that large-scale phenomenology will be predictable via the laws of chance codified in probability theory" (Rescher 2009, p. 107).

In this regard, the interrelation between the epistemological and the ontological realms should be acknowledged, which is especially clear in the case of the scientific prediction. This is because predictability of phenomena or events does not depend only on the nature of those phenomena (the presence of chance, chaos, arbitrary choices, etc.), but it also depends on the questions posed and, therefore, on the type of prediction that we expect to obtain (generic or specific, conditional or categorical, qualitative or quantitative, etc.).

When chance is considered, it might happen that what is actually an epistemological limitation gets mixed up with an ontological obstacle. It is what Rescher calls "fallacy of misattributed chance" (Rescher 1998a, p. 141). It consists of attributing erroneously to chance something that belongs to other factor. This is because, occasionally, we think that our difficulties to predict are due to the presence of chaos, when actually they are due to ignorance (or, potentially, to other factor). This can happen when phenomena at issue are complex and we do not have an appropriate understanding on their operating.

Together with chance, chaos is – in Rescher's judgment – another ontological factor that makes it difficult to predict to a greater extent. He associates chaos with extreme volatility, so that in a chaotic system small differences in its initial state can cause big differences in the result. He considers that "a process if chaotic if *observationally indistinguishable* initial conditions can eventuate in different results, irrespective of how sophisticatedly we make our observations" (Rescher 1998a, pp. 144–145).

In this sense of sensitive dependence of initial conditions, Malanie Mitchell points out that chaos has been observed "in cardiac disorders, turbulence in fluids, electronic circuits, dripping faucets, and many other seemingly unrelated phenomena. These days, the existence of chaotic systems is an accepted fact of science" (Mitchell 2009, p. 20). In effect, chaos can intervene in a great variety of phenomena (natural, social, and artificial). This affects the very possibility of predicting, as well as the level of accuracy and precision that can be achieved by predictions.

In the case of the sciences of nature, chaos theory has pointed out that even a deterministic system can give rise to unpredictable results, since it might be sensitive to variations in the initial conditions (Gonzalez 2015). In this case, the difficulties to predict are due to chaos, instead of being dependent on chance or randomness (Rescher 1998a, p. 143). However, prediction in social sciences is usually more difficult than prediction in the sciences of nature. In this regard, the ontological aspect is crucial, since "it seems clear that the differences in the kind of subject matter have incidence in the role of prediction" (Gonzalez 2015, p. 62).

On the one hand, chaos can intervene in the social systems, just like in the natural phenomena. According to Rescher, "in any system whose workings are subject to a very large number of intricately interacting factors, there is going to be a great sensitivity to parameter determination, so that even a small variation on input values will amplify into substantial variations in output values. This of course is the characteristic situation of chaos (…). Economic systems (and social systems in general) are in large measure chaotic in exactly this sense" (Rescher 1998a, p. 197). Thus, some of the difficulties to social prediction can be due to the presence of chaos.

But, on the other hand, there are other ontological obstacles – besides chaos – that have to do to a greater extent with social phenomena. In this regard, Rescher considers the arbitrary choice as an obstacle to social prediction. In his judgment, "the idea of rationality and its ramifications thus provides us with a very important and useful predictive instrument in the context of human activities" (Rescher 2009, p. 113). Thus, when the behavior of the agents is characterized by being rational, this involves, to some extent, the possibility of anticipating their choices.

From this perspective, it seems clear that irrational behavior sets a limit to predictability about social matters. In effect, if the choices of the agents are arbitrary, it is not possible to anticipate them in a reliable way. However, Rescher considers that arbitrary choice does not involve necessarily unpredictability; because "even where agents behave in erratic and impredictable ways, this need not preclude predictive foreknowledge in the large. For as long as those eccentricities cancel out in the statistical aggregate or become lost in a statistical fog, we can obtain perfectly stable large-scale aggregate effects" (Rescher 2009, p. 114).

However, Rescher considers that, even when the agents' choices are rational, we can have problems to predict them. This is because agents make choices on the basis of the available information. Frequently, this information is not about *how* the things in the world *are*, but about *what* the agents *think* with regard to the reality. This has clear epistemological repercussions, since it involves difficulties to know in an objective way what information the agents use in the decision-making.

Rescher considers that this feature, which goes with many human decisions, allows us to account for the greater difficulty of the social prediction in comparison with the prediction about natural phenomena, as well as its frequent unreliability.[5] In my judgment, the difficulties to social prediction require a broader thematic framework. Thus, they should be analyzed within the *framework of complexity*, instead of relating them just with the ontological or epistemological obstacles to predictability. This is because there is often less complexity in natural systems than in social systems, where there are often more levels at stake.

This greater complexity of the social systems has repercussions on the level of regularity of the phenomena and events that we want to predict. Thus, it is frequent to find regularities in the sciences of nature, where it is possible in many cases to predict in an accurate and precise way due to the possibility of reproduce the phenomena (Gonzalez 2015, p. 61). Meanwhile, when the studied reality is social or artificial, there is always a contextual component that is modulated by historicity and that adds complexity to prediction (Gonzalez 2012b).

From this perspective, when the prediction is about the actions of the agents "one source of this complexity lies in the difficulty in anticipating human behaviour, which can be influenced by a tremendously varied collection of social, political, psychological, biological and other factors. Another source is the inherent difficulty in anticipating the results of interactions of millions of human beings with different values, objectives, motivations, expectations, endowments, rights, means and circumstances, dealing with each other in

5 Cf. Rescher, N., *Personal Communication*, 10.6.2014.

a wide variety of institutional settings" (Sen 1986, p. 5). Undoubtedly, these difficulties due to complexity contribute to the frequent unreliability of the social predictions.

1.3 *Creativity as an Obstacle to Predictability*

When creativity is considered as an obstacle to scientific prediction there are, in principle, two successive levels of analysis: (i) creativity as an element that *configures the object studied* by science (the economic, social, cultural, or politic reality, among others), where agents' creativity can generate changes that involve novelty; and (ii) creativity as a factor that *is part of the very scientific activity* of research (basic or applied) that is relevant both form an structural point of view and from a dynamic perspective.

In the first case – when there is creativity in phenomena or events that are the object of study of science – agents' creativity can be a clear obstacle to scientific prediction (above all, in social sciences and in the sciences of the artificial). In this regard, Rescher relates creativity to unpredictability. Fundamentally, he does this with regard to contexts such as arts, where the presence of creativity allows us to account the difficulties to prediction and even to the impossibility of achieving a prediction (Rescher 2009, p. 103).

When human creativity intervenes, it happens that the pattern followed by a system in the past does not allow us to infer its development in the future (Rescher 2012a, p. 210). For this reason, Rescher considers that "human creativity and inventiveness defies predictive foresight" (Rescher 1998a, p. 149).[6] Even more, one might consider that creativity is a source of complexity. This has been highlighted, above all, in economics, where it is usual to see the creativity of the agents as an obstacle to economic predictions.

Regarding this issue, Wenceslao J. Gonzalez points out that economic agents "frequently develop *creativity* when they perform some actions (e.g., in the design of new financial products). This creativity of the agents adds another element more of complexity while the task is doing economic predictions, because it increases the possibility of novelty in the phenomena instead of [having] the prevalence of the regularity, which is the main ingredient for promoting the reliability of the predictions" (Gonzalez 2012c, p. 147). From this perspective, agents' creativity is another element that gives additional complexity to the economic predictions.

But, together with human creativity *in general*, there is another level of analysis – the second that has been pointed out before – that has to do with

6 Rescher suggests this issue with regard to humanistic realms, as literature, so that he does not consider how it can affect the scientific disciplines related to designs.

scientific creativity, in particular. This is because creativity can be also a feature of the scientific activity; both form a structural point of view and from a dynamic perspective. Nevertheless, it is difficult to characterize what *creativity* consists in, as "the term 'creativity' as commonly used covers too much ground. It refers to very different entities, thus causing a great deal of confusion" (Csikszentmihalyi 1996, p. 25).

This difficulty has been acknowledged by several authors, such as Subrata Dasgupta or Margaret A. Boden. For Dasgupta, there are "many interpretations – some differing in only subtle ways – that may be placed on the very *idea* of creativity" (Dasgupta 1994, p. 16). Meanwhile, Boden has noticed that, in the study of creativity, problems may arise "because of conceptual difficulties in saying what creativity *is*, what *counts* as creative" (Boden 2004, p. 13). In spite of these difficulties, it is usual to accept that the notion of "creativity" involves, in principle, something positive, which is related to something both valuable and novel or original.[7]

In this regard, for the sake of a more rigorous characterization of the notion of "scientific creativity," this concept should be distinguished from the term "innovation." Thus, "creativity" is usually related to a scientific context; while the notion of "innovation" is frequently connected with the realm of technology, where there are often modifications of existing realities rather than something completely new or original (Neira 2012, p. 217). However, "creativity" and "innovation" are interrelated notions, since there are relevant connections between scientific creativity and technological innovation (for instance, in the case of the sciences of design and the information and communication technologies) [Gonzalez 2013a, pp. 11–14].

When scientific creativity is considered in relation to prediction, then creativity can have – in my judgment – a double role: (a) it might be, in principle, an obstácle for predicting the future science; and (b) it can be a key factor for problem-solving, so it can contribute to overcome the current unpredictability of the phenomena and events. Usually, Rescher's attention is focused in the former case: creativity as an impediment to the prediction about the development of the future science (Rescher 2012b). But he offers a thematic framework that makes it also possible to analyze creativity as a medium to deal with the obstacles to predictability (above all, in the epistemological and methodological levels).

Regarding the first role (creativity as an obstacle to the prediction of the future science), Rescher focuses his attention in one of the aspects of scientific creativity: conceptual creativity. This is because, within a framework of

7 On the notion of creativity, see Gonzalez (2013a).

pragmatic idealism, concepts are crucial for the articulation of knowledge. In this way, his thought gives priority to the epistemological realm, as a way to address the problems posed. This also happens when he considers scientific creativity. Thus, he thinks that science is open to conceptual creativity, so changes occur regarding the scientific concepts.[8]

Conceptual creativity is a key feature to address scientific progress and the problem of the limits of science, since research is always carried through from a conceptual framework open to changes. So it happens that taking a concrete conceptual framework sets, in practice, limits to scientific research, insofar as "the major discoveries of later stages are ones which the workers of a substantially earlier period (however clever) not only have failed to make but which they could not even have understood, because the requisite concepts were simply not available to them" (Rescher 2012b, p. 151). Thus, conceptual creativity affects the predictive ability of science, since certainly we cannot predict those things that we cannot conceive (Rescher 2012b, pp. 149–163).

However, conceptual creativity is – in my judgment – an aspect *among others* of scientific creativity. In effect, from a structural point of view, creativity can be considered in relation to the constitutive elements of science: language, structure, knowledge, method, activity, ends, and values.[9] Therefore, there might be creativity in scientific language (through the introduction of new terms or through changes of meaning in the existing terms), in the structure of scientific theories (where there are alternatives to the dominant theory), in scientific knowledge (through new concepts and new problems), in methods (where there are changes in the procedures), with regard to scientific activity (that involves novelty regarding the aims, processes, and results of the research) and in scientific values (internal and external) that modulate the selection of ends (Gonzalez 2013a, pp. 15–17).

From this perspective, scientific creativity might involve the possible "not predictable" situation or even the "unpredictability" about the development of future science in the diverse elements pointed out (language, structure, knowledge, method, activity, ends, and values), instead of being something due only to changes regarding the concepts. At the same time, it might be noticed that the problems to predict the future science, which are mainly the result of the presence of creativity, depend to a large extent on the time horizon of the scientific prediction. Thus, it seems possible to anticipate with enough reliability the aims, processes, and results of science in the short run; while it could be certainly impossible to predict them in the long run (Gonzalez, W.J. 2016).

8 On conceptual change, see Gonzalez (2011b, pp. 39–62).

9 On the constitutive elements of science, see Gonzalez (2005, pp. 3–49; especially, pp. 10–11).

It happens that scientific creativity, besides being an important obstacle to the predictability of future science, can be also a medium to overcome the limits of science, in general, and the limits to scientific prediction, in particular. This can happen both in a context of basic science, where creativity connects above all with explanation and prediction, and in the realm of the applied science, where creativity is mainly related to prediction and prescription. In addition, creativity can be considered in relation to the application of science, which has to do with the use of scientific knowledge by agents, so creativity can lead to new applications of the available knowledge (Gonzalez 2013a, pp. 17–18).

Therefore, while human creativity, in general, is a clear obstacle to scientific prediction (for example, in economics, insofar as it favors a high degree of novelty), scientific creativity can be dual in this regard. On the one hand, creativity is certainly an obstacle that makes it difficult to predict the future science, above all when the prediction is in the long run. But, on the other hand, it can contribute to overcome the limits to predictability, because it allows us to open new routes, regarding explanation and prediction in basic science as well as with regard to prediction and prescription in applied science.

Clearly, the limits to the task of predicting – among them, the ontological limits – make it impossible to achieve perfection in the predictive domain. Taken them together, the ontological limitations to predictability involve the impossibility of achieving what Rescher calls "predictive completeness." This includes that science could accurately predict those things that are in principle recognized as predictable by science itself (Rescher 1999, pp. 137–148). Thus, "only the course of experience can inform us about the extent to which the phenomena of a particular domain are predictable. And with predictability in general, just as with specific issues of prediction, one must simply wait and see" (Rescher 1998a, pp. 81–82).

2 The Ontological Obstacles to Scientific Prediction from the Perspective of Complexity

Ontological limits to predictability (volatility, chance, chaos, etc.) can be seen as sources of complexity for predictions. Certainly, the perspective of complexity is relevant when the limits of scientific prediction are analyzed from an ontological point of view. This is because prediction can be about a reality (natural, social, or artificial) that can be characterized as complex. Thus, its complexity has repercussions that go in two directions: on the one hand, it has repercussions on the very possibility of predicting; and, on the other, it modulates the configuration of the predictive processes and the type of prediction that can be achieved (its reliability, accuracy, precision, etc.).[10] In this regard, it should be taken into account that complexity is a twofold notion: it has a structural dimension and a dynamic trait (Gonzalez 2013b, pp. 299–311).

The structural dimension of complexity has to do with the parts that compound the complex system and the interactions between those parts. Meanwhile, the dynamic trait of complexity has to do with the changes that occur in the system over the time, both from an internal point of view (usually in terms of aims, processes, and results) and from an external perspective (that deals with the relations with the context or environment). Usually, Rescher pays more attention to the structural dimension of complexity than to the dynamic trait. A broader approach to complexity and its repercussions on the limits of predictability must also pay attention – in my judgment – to the complex dynamics ("internal" and "external"), where historicity can be a key aspect, above all when the reality researched is a social or artificial reality.

2.1 *Varieties and Modes of Complexity*

When the complexity of a system (natural, social, or artificial) is considered, two different dimensions of complexity should be taken into account in principle: (a) the structural dimension (that has to do with the parts that compound the system and their interactions); and (b) the dynamic trait (that deals with the change over the time in the system). Frequently, the studies of complexity are focused on the structural perspective. Even more, there are characterizations of complexity, in general, that only address the complex structure, without taking into account, in fact, the dynamic complexity.[11]

An interesting and especially influential approach to complexity is that offered by Herbert A. Simon, centered in the sciences of the artificial, in general, and the sciences of design, in particular. His focus is on "organized complexity," instead of paying attention to a kind of complexity that is disorganized (for example, a chaotic structure). From this perspective, Simon considers that designs consist of a series of elements that are coordinated, and they may have

10 It has been noticed that complexity sets limits to science: "the study of the characteristics of complex dynamic systems are showing us exactly why limited knowledge is unavoidable – or, to be more precise, why knowledge *has to be* limited," Cilliers (2007, p. 82).

11 "Complexity depends on the number of elements of the system, the number of its properties and the number of relationships between these elements or properties." Betz (2006, p. 81). This characterization can be valid with regard to the complex structure of a system, but it does not encompass – in my judgment – the whole realm of complexity, since it does not take into account the dynamics.

a hierarchic configuration, in such a way that complex systems are nearly-decomposable. Thus, their components cannot be completely separated, since they are interdependent (Simon 1996, Ch. 8, pp. 183–216).

In this way, Simon considers that, in scientific research, *analysis* is the complement of the task of synthesis carried out by design sciences (Simon 1995, p. 246). From the structural perspective, his conception of the analysis is commonly *hological*, and it is oriented towards nearly-decomposability. So, when the aim is to clarify the structural complexity of design sciences, he thinks that a hological analysis is required. This is so because "the complexity of a whole should be decomposed (...), where there are numerous parts that interact. When this operation is carried out, insofar as the subsystems that compound the complex system are inter-dependent, it is not possible that the subsystems can split up completely ones from others" (Gonzalez 2012a, p. 17).

His view of a complex system is based mainly on a hological perspective: the relations between the parts and the whole are emphasized, although he avoids giving a precise definition of "complex systems."[12] Generally, he considers that a complex system is "one made up of a large number of parts that have many interactions" (Simon 1996, pp. 183–184). In these systems, the whole is more than the sum of the parts, in such a way that "given the properties of the parts and the laws of their interaction, it is not a trivial matter to infer the properties of the whole" (Simon 1996, p. 184). Thus, in a broad sense, complexity can be understood as opposite to simplicity (2001, pp. 32–72).

In my judgment, such an approach to complexity is too general. On the one hand, structural complexity cannot be only an organized complexity, but it can also be found disorganized complexity in complex systems (for example, in chaotic systems). At the same time, the notion of "hierarchy" is not good enough to undertake the whole organized complexity, since it is also possible to have a "poli-hierarchy." In this respect, the need for "coordination" has been highlighted (Rescher 1998b, pp. 10–11).[13] On the other hand, Simon's approach to complexity is too focused on the structural perspective.[14] In this regard, in order to grasp the whole complexity, it seems necessary a deeper conception

12 It must been pointed out that there is not a definition of "complexity" or "complex system" that is generally accepted. Cf. Chu, Strand and Fjelland (2003, p. 19).

13 A comparison between Rescher and Simon's approaches to complexity is offered in Gonzalez (2012b, pp. 76–79 and 2012c, pp. 151–153).

14 Certainly, the dynamic trait is also present in Simon's analysis of the sciences of design as sciences of complexity, but he is more interested in the structure of the complex systems than in the dynamic complexity. Cf. Simon (1996).

open to dynamic complexity, both of science by itself and of the studied reality (above all, when it is a social or an artificial reality).

Rescher's approach to complexity goes further – in my judgment – than Simon's proposal. In effect, Rescher considers that Simon's conception on complexity is too generic, since there are few real things that can escape complexity when it is understood in terms of the relations between the parts and the whole (Rescher 1998b, p. 22, n. 44). Thus, although Rescher is also mainly focused on "structural complexity," he highlights that there are numerous features of complexity. He does this through a breakdown of the modes of complexity according with two basic levels: the epistemic and the ontological ones.

Epistemic complexity – as Rescher sees it – is related to formulas. Within this first level of complexity, he proposes three epistemic or cognitive modes of complexity: (i) descriptive complexity, which deals with the level of difficulty that involves describing a system in a proper way; (ii) generative complexity, which reflects the number of steps that are required to give rise to the complex system; and (iii) computational complexity, which is associated with the number of resources required in order to solve problems related to the system (Rescher 1998b, p. 9).

There is another level of complexity in Rescher, which is of an *ontological* kind (the complexity in reality itself). In this regard, he distinguishes three ontological modes of complexity. In turn, each one of them has several possible options: (a) compositional complexity (constitutional and taxonomic); (b) structural complexity (organizational and hierarchical); and (c) functional complexity (operational and nomic) [Rescher 1998b, p. 9]. With this collection of the epistemic and ontological modes of complexity, Rescher offers a framework of complexity that is certainly more systematic and broader than other approaches.

His analysis of ontological complexity can be used to address the problem of scientific prediction and its limits. In this regard, when scientific prediction is analyzed from an ontological perspective, Rescher's proposal on the ontological modes of complexity certainly has philosophical relevance. This is because the reality predicted as well as the reality used for prediction could be a complex reality. So the study of this ontological complexity (that has repercussions on the epistemological and methodological levels) can have a key role to characterize the prediction to be made. On the one hand, it allows us to report some obstacles to predictability; and, on the other, it can help to overcome those obstacles.

Thus, Rescher's proposal on complexity is really interesting, but it is excessively focused on the structural complexity and he scarcely pays attention to the dynamic trait. A broader approach to complexity should – in my judgment – pay attention to dynamic complexity too, that is connected with the notion of historicity. This is especially clear when scientific prediction is considered, since it is oriented towards a possible future, so it has to deal with the changes that occur in the systems over time.

2.2 *Ontological Modes of Complexity and Their Implications for the Limits to Predictability*

A system can be complex due to its own ontological characteristics. This issue underlies Rescher's identification of the "modes of complexity." His analysis is focused on the epistemic modes of complexity (those related to knowledge) and the ontological modes of complexity (that are related to reality itself). The *methodological modes* of complexity can be added to his proposal, since the complexity related to reality and to knowledge has repercussions on the scientific processes (for example, in the case of the predictive models).

All these modes of complexity (epistemic, ontological, and methodological) have repercussions on science, in general, and on scientific prediction, in particular. In this regard, within a pragmatic idealism such as that defended by Rescher, ontological complexity is especially relevant, insofar as it is the support for the epistemological level (the achievable knowledge) and the methodological side (the scientific processes). From an ontological perspective – which has to do with the reality itself researched (natural, social, or artificial) – a system can be complex (for him) in accordance with three modes of complexity: (i) compositional complexity; (ii) structural complexity; and (iii) functional complexity (Rescher 1998b, p. 9).

Compositional complexity is related to the elements that compound the system and the variety of those elements. Thus, Rescher distinguishes two forms of complexity related to system's composition: *constitutional complexity* and *taxonomic complexity* or *heterogeneity*. Constitutional complexity is associated with the number of elements that compound a system, while taxonomic complexity or heterogeneity has to do with the variety of those elements (Rescher 1998b, p. 11).

Structural complexity has in Rescher a restricted sense, since it is circumscribed to the organization and relation between the elements of the system, instead of encompassing all the structural factors of complexity. Within what he calls "structural complexity," there are also two possibilities: *organizational complexity* and *hierarchical complexity*. The former is related to the forms of organization that can be seen in the system (i.e., it has to do with the possible forms of interrelations between its components). The later has to do with the possible ways to relate the elements of the system (Rescher 1998b, pp. 11–12).

Meanwhile, functional complexity is related to the behavior of the system that clearly involves a temporal dimension. It is diversified, in turn, in two modes of complexity: *operational complexity* and *nomic complexity*. Operational complexity studies the variety of the modes of operation and the types of functioning that appear in the system, an issue that he sees in terms of processes. Nomic complexity is related to the patterns that regulate the relations between the elements of the system (Rescher 1998b, pp. 12–13). In this regard, ontological obstacles to predictability such as anarchy, volatility, or chaos can be seen as sources of functional complexity.

Through this framework about the ontological complexity (that pays attention to three features: the composition, the structure, and the function of the complex systems) Rescher highlights that, when we try to encompass the complex reality of a complex system (natural, social, or artificial) there are a wide variety of factors to take into account. Thus, the ontological complexity of a system involves that we have to take into account the components (their number and variety), the structure (regarding the organization and the relations between the components), and the functions (the types of operation and the working patterns).

From this perspective, it can be clearly seen that prediction of a complex system with ontological complexity has to deal with a wide variety of factors. This is so because, although a system does not need to be complex in all the aforementioned senses, "the different modes of complexity do tend to run together in practice" (Rescher 1998b, p. 15). Thus, when a system is complex with regard to its composition and structure, it is usually complex regarding their function too (1998b, p. 15). Certainly, this makes it difficult to predict, and has repercussions on issues such as the reliability or the degree of accuracy and precision that can be achievable.

But it happens that, regarding complex systems, there are still more factors to take into account than only the epistemic and ontological modes of complexity. Thus, although Rescher's approach to complexity is really detailed, his attention is mainly on the structural dimension of complexity (understood in a broader sense, which also encompass the compositional complexity). For that reason, although his approach is open to the dynamic trait (that has to do with the changes that occur in the system over the time) it is a perspective that, in my judgment, is not exhaustively developed in Rescher.

In effect, when dynamic complexity is considered, we must take into account, initially, two dimensions: the "internal" dynamics and the "external" dynamics (Gonzalez 2012a, pp. 13–14). The internal dynamics is related to the activity of the system itself, whereas the external dynamics deals with the relations to the environment. Rescher pays special attention to the "internal" dynamics of the complex system, which he considers in accordance with the notion of "process."[15] But he does not pay attention *de facto* to the "external" dynamics, where there are sources of complexity regarding the relations of the system with the context.

In this regard, the concept of "historicity" has been proposed to characterize the change in the complex dynamics ("internal" and "external") and its repercussions on scientific prediction (Gonzalez 2012b, pp. 79–88). It is a notion that goes further than the mere temporality, since it is difficult to think of a reality that escapes from the temporal dimension. However, historicity is a characteristic of the human realm: it links with the human activity that is *eo ipso* historic, in the sense that it changes over time, and some changes are particularly relevant (Gonzalez 2003, p. 88). In addition, historicity has into account the context that surrounds human activity.

Thus, historicity is related to three levels of analysis: science, the agents, and the reality itself that is researched. This triple level is pointed out by Wenceslao J. Gonzalez: "(1) Historicity (*Geschichtlichkeit*/*historicidad*) is a trait of science, in general, and each science, in particular. This facet can be found in the whole set of constitutive elements of science, such as language, structure, knowledge, method, activity, ends, and values. (2) Historicity configures the agents themselves involved in the development of scientific research, insofar as they are human beings within a historical context. (3) Historicity is a characteristic of the reality itself that is researched (above all, in the social and artificial realms)" (Gonzalez 2011b, p. 43).

When these three levels of analysis are considered, it seems clear that the notion of *historicity* is relevant when the problems of complexity, in general, and its repercussions on scientific prediction, in particular, are considered. In the first place, historicity is a feature that modulates science. This happens in three successive levels: the constitutive elements of science (that lead to the structural dimension of complexity), the configuration of the aims, processes, and results (the internal dynamics of scientific activity), and the relations with the changing environment (the external dynamics).

In the second place, historicity configures the agents who develop the scientific activity, who are in a concrete socio-historical context. This feature can add more complexity (for instance, through human creativity, both individual and social). Finally, historicity can be a feature of the reality itself studied by science, above all when it is a social or artificial reality. From this perspective, historicity can be the basis to understand why prediction is more

15 On the notion of "process" in Rescher, see Gonzalez (2012b, pp. 79–80).

difficult in social sciences and the sciences of the artificial, where the complex systems are historical.

From this perspective, it seems to me that the ontological obstacles to scientific prediction can be better understood when they are related to a broader framework of complexity (both structural and dynamic). Moreover, the dynamic trait of complexity (internal and external) is especially relevant, since the ontological obstacles to predictability that have been analyzed (anarchy, volatility, chance, chaos, arbitrary choice, and creativity) have to do with changes over time that set limits to rational prediction. In addition, a thoughtful account of complexity should take into account the notion of "historicity," which is required when issues such as agent's choices or creativity (human and scientific) are considered.

3 Obstacles to Predictability from the Dynamic Trait: Complexity and Historicity

A key issue for the reflection on the ontological limits to predictability in the social sciences and the sciences of the artificial is the problem of complexity. Even more, these sciences can be analyzed as sciences of complexity, both with respect to their structure and with regard to their dynamics (internal and external). In effect, they are sciences with structural complexity. This can be seen in relation to their constitutive elements, such as language, internal articulation, knowledge, method, activity, ends, and values (Gonzalez 2012b, p. 74). Additionally, they have a complex dynamics (internal and external), above all when they are configured as applied sciences (Gonzalez 2012a, pp. 8–9). Thus, in their internal dynamics, there is an articulation of aims and processes, which generate a result. At the same time, they involve an external dynamics, since their activity is developed in a changing context.

Both the structural dimension and the dynamic trait of complexity in social sciences and the sciences of the artificial are modulated by historicity. This is a feature that they have in common with the natural sciences, since every science is the result of the human activity, which is historic in itself. But historicity is also an ontological feature of the reality researched by social sciences and the sciences of the artificial, since they are disciplines that study the social reality and the human-made realm, respectively. The social ontology and the artificial ontology are both under the influence of historicity.

From this perspective, it seems clear that there is also a contextual component in the scientific undertakings. "Because the agents – individual and social – act usually with regard to contexts (historical, cultural, political, ...), which are changing" (Gonzalez 2012a, p. 14). So historicity allows us to understand the huge variability of the reality (above all, social and artificial), because it offers the appropriate framework to address factors like human creativity or the choices of the agents that are a source of complexity for predictions in social sciences and sciences of the artificial.

It is usual to consider that prediction in social sciences is more difficult than in natural sciences, both in terms of possibility and in terms of results. At the same time, it is also usual to think that this difficulty of social prediction is because there is more complexity in the social systems (or in the artificial systems) than in the natural systems (Gonzalez 2011a, pp. 319–330; especially, pp. 319–321). Thus, from a structural point of view, more factors might intervene, with more complex interactions; and, from a dynamic perspective, the variability is usually higher.

However, the studies on complexity, both in social sciences and in the sciences of the artificial, do not take usually into account historicity. Whereas, they are focused in notions as "evolution" or "process," which do not achieve the depth that historicity achieves (Gonzalez 2012b, pp. 79–84). In my judgment, a more sophisticated analysis of complexity (structural and dynamic) can be developed through historicity, above all in the social and artificial realms. In the case of economics, this has been highlighted by Wenceslao J. Gonzalez. Economics is a dual science (it is a social science and also a science of the artificial), so it involves features of both groups of disciplines.

Gonzalez considers that, when the problem of complexity in economics and its repercussions on economic prediction is analyzed, there is an ontological duality that must be taken into account in its configuration. This ontological duality is the differentiation between "economic activity" and "economics as activity" (Gonzalez 1994, pp. 253–294). It is a distinction that involves a relevant source of complexity for economics, as a human endeavor imbued with historicity. This complexity adds difficulty to economic prediction, which has to take into account the duality of components "economic activity" and "economics as activity."

On the one hand, *economic activity* can be understood, in principle, as autonomous regarding other human activities, since it has its own characteristics. Thus, it involves the activities carried through by human beings, which include supply and demand, the transactions of good and services, etc. On the other hand, *economics as activity* is a human endeavor that is related with other human activities (cultural, political, social, ecological, etc.). Commonly, it is interconnected with them in many ways. From this perspective, there are relations with the context, so that economics as human activity – which is interrelated with other activities – is not independent of the context (Gonzalez 2012b, pp. 86–87).

It happens that, through economic activity, "prediction has to deal with forms of complexity that are genuinely of an economic character" (Gonzalez 2012c, p. 146). At the same time, economics as activity "involves that scientific prediction should consider a series of different types of complexity (social, political, cultural, ecological, etc.), that certainly have repercussions on the economic variables" (Gonzalez 2012c, p. 146). In both cases, historicity is a key feature, since economic reality cannot be understood without taking into account the historicity of human activity, developed in a social context.

Therefore, the duality "economic activity" and "economics as activity" involves that "the predictability of economic activity – which is, in principle, autonomous – is possible, and the prediction could be reliable; whereas predictability of economics as a human activity among others appears more unreliable, precisely due to the interdependence with other activities" (Gonzalez 2012b, p. 87). But accepting the historicity of the economic reality does not involve, certainly, neglecting its objectivity. Both "economic activity" and "economics as activity" are objective realities. However, whereas economic activity is, in principle, measurable, economics as activity is more difficult to measure, so it is not usually included in the neoclassical econometrical models (Gonzalez 2012b, p. 87).

With regard to the difficulties to predict in social sciences, in general, and in economics, in particular, Rescher insists on the fact that those difficulties are mainly rooted in the presence of cognitive factors. This is because social prediction, *prima facie*, is about the acts and choices of the agents, and it happens that agents do not make decisions on the basis of the reality of the world, but they make decisions on the basis of what they *think* of that reality. In this way, the acts and choices of the agents depend on factors such as their beliefs, ideas, etc., that are variable.[16] This means that there is a subjective component, and social prediction has to deal with it.

However, if the distinction "economic activity" and "economics as activity" is taken into account, it seems that there are even more sources of complexity that affect economic prediction, so there are more factors than those contemplated by Rescher. This situation, in my judgment, allows us to understand the higher difficulty of economic prediction in comparison with prediction in the sciences of nature (for example, in physics). In addition, this distinction – that

16 Cf. Rescher, N., *Personal Communication*, 10.6.2014.

has as it starting point the acknowledgement of the historicity that configures the social and artificial reality – can be applied to other dual sciences, such as the communication sciences (Arrojo 2012).

4 Coda: Towards a Broad Approach to the Limits of Predictability from the Ontology of Science

Regarding the limits of scientific prediction, Rescher has identified and analyzed several of the most relevant ontological obstacles to predictability: anarchy, volatility, chance, chaos, arbitrary choice, and creativity. It is an especially relevant issue, since the reality of phenomena has methodological repercussions (the predictive models have to adapt to the reality that these models want to predict) and epistemological incidence (it affects the kind of knowledge that can be achieved). In this regard, the ontological obstacles to predictability can be seen as a source of complexity for predictions.

From this perspective, a broad approach to complexity is required in order to provide a cogent account of the limits to predictability, in general, and the ontological limits, in particular. Although Rescher has into account many of the main ontological limits to predictability, his proposal regarding complexity in science is – in my judgment – too focused on the structural dimension. But it happens that the ontological obstacles to predictability (anarchy, volatility, chance, chaos, etc.) have to do with a temporal dimension, insofar as they are related with changes that might occur in phenomena over time. Moreover, some of them, such as creativity and agent's choices, have a clear historical component, that should be acknowledged in order provide a broader analysis of those limits.

To sum up, Rescher's approach to the epistemic and ontological modes of complexity allows us to understand some of the methodological difficulties to prediction (and, consequently, some limits of predictability). But his approach should be incorporated into a broader account of complexity, where the ontological obstacles to predictability can be encompassed. Thus, we have to consider two aspects of complexity: the structural dimension and the dynamic trait (internal and external). In order to do this, it seems to me that historicity should be explicitly acknowledged, since it is a key feature of science, the agents, and the reality itself that is researched (above all, social and artificial). Furthermore, it provides the appropriate framework to address some of the sources of complexity for predictions in social sciences and the sciences of the artificial, such as human creativity and the choices of the agents.

References

Arrojo, M.J. 2012. "La sobriedad de factores en el análisis de la complejidad en las Ciencias de la Comunicación. El estudio de la Televisión Digital Terrestre," in Gonzalez ed. *Las Ciencias de la Complejidad: Vertiente dinámica de las Ciencias de Diseño y sobriedad de factores*, (pp. 337–358). A Coruña: Netbiblo.

Betz, G. 2006. *Prediction or Prophecy? The Boundaries of Economic Foreknowledge and Their Socio-Political Consequences*. Wiesbaden: Deutscher Universitäts-Verlag.

Boden, M.A. 2004. *The Creative Mind. Myths and Mechanisms*, 2nd ed. London: Routledge (1st ed. 1990).

Chu, D., Strand, R. and Fjelland, R. 2003. "Theories of Complexity. Common Denominators of Complex Systems," in *Complexity*, 8 (3), 19–30.

Cilliers, P. 2007. "Why We Cannot Know Complex Things Completely," in Capra, F., Juarrero, A., Sotolongo, P., and van Uden, J. eds. *Reframing Complexity. Perspectives from the North and South, Exploring Complexity*, vol 1., (pp. 81–89). Mansfield, MA: ISCE.

Csikszentmihalyi, M. 1996. *Creativity. Flow and the Psychology of Discovery and Invention*. N. York: Harper/Collins, (reprinted in 1997).

Dasgupta, S. 1994. *Creativity in Invention and Design*. Cambridge: Cambridge University Press.

Gonzalez, W.J. 1994. "Economic Prediction and Human Activity. An Analysis of Prediction in Economics from Action Theory," in *Epistemologia*, 17 (2), 253–294.

Gonzalez, W.J. 2003. "Racionalidad y Economía: De la racionalidad de la Economía como Ciencia a la racionalidad de los agentes económicos," in Gonzalez, W.J. ed. *Racionalidad, historicidad y predicción en Herbert A. Simon*, (pp. 65–96). A Coruña: Netbiblo.

Gonzalez, W.J. 2005. "The Philosophical Approach to Science, Technology and Society," in Gonzalez, W.J. ed. *Science, Technology and Society: A Philosophical Perspective*, (pp. 3–49). A Coruña: Netbiblo.

Gonzalez, W.J. 2010. *La predicción científica. Concepciones filosófico-metodológicas desde H. Reichenbach a N. Rescher*. Barcelona: Montesinos.

Gonzalez, W.J. 2011a. "Complexity in Economics and Prediction: The Role of Parsimonious Factors," in Dieks, D., Gonzalez, W.J., Hartman, S., Uebel, Th. and Weber, M. eds. *Explanation, Prediction, and Confirmation*, (pp. 319–330). Dordrecht: Springer.

Gonzalez, W.J. 2011b. "Conceptual Changes and Scientific Diversity: The Role of Historicity," in Gonzalez, W.J. ed. *Conceptual Revolutions: From Cognitive Science to Medicine*, (pp. 39–62). A Coruña: Netbiblo.

Gonzalez, W.J. 2012a. "Las Ciencias de Diseño en cuanto Ciencias de la Complejidad: Análisis de la Economía, Documentación y Comunicación," in Gonzalez, W.J. ed. *Las Ciencias de la Complejidad: Vertiente dinámica de las Ciencias de Diseño y sobriedad de factores*, (pp. 7–30). A Coruña: Netbiblo.

Gonzalez, W.J. 2012b. "La vertiente dinámica de las Ciencias de la Complejidad. Repercusión de la historicidad para la predicción científica en las Ciencias de Diseño," in Gonzalez, W.J. ed. *Las Ciencias de la Complejidad: Vertiente dinámica de las Ciencias de Diseño y sobriedad de factores*, (pp. 73–106). A Coruña: Netbiblo.

Gonzalez, W.J. 2012c. "Complejidad estructural en Ciencias de Diseño y su incidencia en la predicción científica: El papel de la sobriedad de factores (*parsimonious factors*)," in Gonzalez, W.J. ed. *Las Ciencias de la Complejidad: Vertiente dinámica de las Ciencias de Diseño y sobriedad de factores*, (pp. 143–167). A Coruña: Netbiblo.

Gonzalez, W.J. 2013a. "The Roles of Scientific Creativity and Technological Innovation in the Context of Complexity of Science," in Gonzalez, W.J. ed. *Creativity, Innovation, and Complexity in Science*, (pp. 11–40). A Coruña: Netbiblo.

Gonzalez, W.J. 2013b. "The Sciences of Design as Sciences of Complexity: The Dynamic Trait," in Andersen, H., Dieks, D., Gonzalez, W.J., Uebel, Th. and Wheeler, G. eds. *New Challenges to Philosophy of Science*, (pp. 299–311). Dordrecht: Springer.

Gonzalez, W.J. 2015. *Philosophico-Methodological Analysis of Prediction and its Role in Economics*. Dordrecht: Springer.

Gonzalez, W.J. 2016. "Rethinking the Limits of Science: From the Difficulties for the Frontiers to the Concern on the Confines," in Gonzalez, W.J. ed. *The Limits of Science: An Analysis from "Barriers" to "Confines,"* (pp. 3–30). Brill.

Granger, C.W.J. 1989. *Forecasting in Business and Economics*, 2nd ed. San Diego, CA: Academic Press, (1st ed., 1980).

Mitchell, M. 2009. *Complexity: A Guided Tour*. Oxford: Oxford University Press.

Neira, P. 2012. "Complejidad en Ciencias de la Comunicación debido a la racionalidad: Papel de la racionalidad limitada ante la creatividad e innovación en Internet," in Gonzalez, W.J. ed. *Las Ciencias de la Complejidad: Vertiente dinámica de las Ciencias de Diseño y sobriedad de factores*, (pp. 205–230). A Coruña: Netbiblo.

Rescher, N. 1995. "Predictive Incapacity and Rational Decision," in *European Review*, 3 (4), 327–332.

Rescher, N. 1998a. *Predicting the Future. An Introduction to the Theory of Forecasting*. N. York: State University of New York Press.

Rescher, N. 1998b. *Complexity: A Philosophical Overview*. N. Brunswick, NJ: Transaction Publishers.

Rescher, N. 1999. *Razón y valores en la Era científico-tecnológica*. Barcelona: Paidós.

Rescher, N. 2009. *Ignorance. On the Wider Implications of Deficient Knowledge*. Pittsburgh, PA: University of Pittsburgh Press.

Rescher, N. 2012a. "Political Pragmatism," in Rescher, N., *Pragmatism. The Restoration of its Scientific Roots*, (pp. 205–215). N. Brunswick, NJ: Transaction Publishers.

Rescher, N. 2012b. "The Problem of Future Knowledge," *Mind and Society*, 11 (2), 149–163.

Sen, A. 1986. "Prediction and Economic Theory," in Mason, J. Mathias, P. and Westcott, J.H. eds. *Predictability in Science and Society*. London: The Royal Society and The British Academy.

Simon, H.A. 1995. "Problem Forming, Problem Finding, and Problem Solving in Design," in Collen, A. and Gasparski, W.W. eds. *Design and Systems: General Applications of Methodology*, v. 3, (pp. 245–257). New Brunswick, NJ: Transaction Publishers.

Simon, H.A. 1996. *The Sciences of the Artificial*, 3rd ed. Cambridge, MA: The MIT Press. (1st ed., 1969; 2nd ed., 1981).

Simon, H.A. 2001. "Science Seeks Parsimony, not Simplicity: Searching for Pattern in Phenomena," in Zellner, A., Keuzenkamp, H.A., and McAleer, M. eds. *Simplicity, Inference and Modelling. Keeping It Sophisticatedly Simple*, (pp. 32–72). Cambridge: Cambridge University Press.

Index of Names

Subject Index

www.ingramcontent.com/pod-product-compliance
Lightning Source LLC
LaVergne TN
LVHW010626300926
849162LV00001B/80
9789004325395